Q. E. D., or New Light on the Doctrine of Creation

by George McCready Price

Preface

The great world disaster, ushered in with the dawn of that August morning in 1914, has already brought revolutionary changes in many departments of our thinking. But not the least of the surprises awaiting an amazed world, whenever attention can again be directed to such subjects, will be the realization that we have now definitely outgrown many notions in science and philosophy which in the old order of things were supposed to have been eternally settled.

There are but two theories regarding the origin of our world and of the various forms of plants and animals upon it, Creation and Evolution,--the latter assuming many modifications.

The essential idea of the Evolution theory is _uniformity_; that is, it seeks to show that life in all its various forms and manifestations probably originated by causes similar to or identical with forces and processes now prevailing. It teaches the absolute supremacy and the past continuity of natural law as now observed. It says that the changes now going on in our modern world have always been in action and that these present-day natural changes and processes are as much a part of the origin of things as anything that ever took place in the past. In short, Evolution as a philosophy of nature is an effort to smooth out all distinction between Creation and the ordinary processes of nature that are now under the regime of "natural law."

On the other hand, the essential idea of the doctrine of Creation is that, back at a period called the "beginning," forces and powers were brought into exercise and results were accomplished that have not since been exercised or accomplished. That is, the origin of the first organic forms, indeed of the whole world as we know it, was essentially and radically different from the ways in which these forms are perpetuated and the world sustained to-day. Time is in no way the essential idea in the problem. The question of how much time was occupied in the work of Creation is of no importance, neither is the question of how long ago it took place. The one essential idea is that in its nature Creation is essentially inscrutable; we can never hope to know just how it was accomplished; we cannot expect to know the process or the details, for we have nothing with which to measure it. The one essential thing in the doctrine of Creation is that the origin of our world and of the things upon it came about at some period of time in the past by a direct and unusual manifestation of Divine power; and that since this original Creation other and different forces and powers have prevailed to sustain and perpetuate the forms of life and indeed the entire world as then called into existence.

Accordingly, we might establish the Evolution doctrine by showing that matter can be made _de novo_, that energy can be created or increased in amount, that life can be made from the not-living, and that new and distinct forms of life can be produced in modern times,--all by natural law as now prevailing.

Or we can practically demonstrate the historical reality of a direct Creation at some time in the past, if we can show that the net results of all modern science tend to prove that the forces and processes now in operation can never account for the origin of things; that matter, and energy, and life, and the various forms of life must all have had an origin essentially different from anything now going on around us.

This indicates the line of argument adopted in the following pages.

The Evolution theory has been widely discussed and accepted in modern times. Indeed it has had a fair chance and an open field for several decades. What is the present situation of the controversy? The friends of the Bible and of old-fashioned Christianity need to know the real facts of the present situation.

Every now and then the news despatches report that the great Professor So-and-so has at last really produced life from the not-living, or has obtained some absolutely new type of life by some wonderful feat of breeding. Or some geologist or archeologist has discovered in the earth the missing link which connects the higher forms of life with the lower, or which bridges over the gulf between man and the apes. Thus many people who get their "science" through the daily papers really believe that these long-looked-for proofs of Evolution have at last been demonstrated, and hence they receive without question the confident assertions of the camp followers of science published at space rates in the Sunday supplements that all intelligent men of to-day have long ago accepted the Evolution doctrine.

But in spite of the quick dissemination of news and the universal spread of education, it seems but a slow process for the really important discoveries of modern science to filter down through such media as the current periodicals to the rank and file of society. The situation seems to illustrate the old adage that a lie will travel round the world while truth is getting on her shoes. _Thus it happens that the common people are still being taught in this second decade of the twentieth century many things that real scientists outgrew nearly a generation ago, and assertions are still being bandied around in the individual sciences which are wholly unwarranted by a general survey of the whole field of modern natural science_. Indeed, in almost every one of the separate sciences the arguments upon which the theory of Evolution gained its popularity a generation or so ago are now known by the various specialists to have been blunders, or mistakes, or hasty conclusions of one kind or another. Thus the market value of all the various subsidiary stocks of the Evolution group has been steadily declining in their respective home markets, and now stands away below par; while strange to say the stock of the central holding company itself is still quoted at fictitiously high figures.

This curious--not to say deplorable--situation has developed largely because of the modern system of strict specialization in the various departments of science. Each scientist feels compelled by an unwritten but rigid code of professional ethics to confine himself strictly to the cultivation of the little plot of ground on which he happens to be working, and is forbidden to express an opinion about what he may know has been discovered on another plot of ground on which his neighbor is working, except by express permission. In other words, science teaching has now become strictly a matter of authority, this authority being vested in the various specialists; and nobody is permitted to look at it in a broad way, or to frame a general induction from the sum of all the facts of nature now discovered, under penalty of scientific excommunication. The scientific code of ethics forbids any general view of the woods: each man must confine himself to the observation of the particular tree in front of his own nose.

But these pages have been prepared under the idea that it is high time to take a more general survey of the geography, time to take our eyes off the various individual trees, and to look at the woods. Perhaps in some respects they may be regarded as too technical for ordinary readers. But if this is the case, it is because the writer had to choose between this somewhat technical treatment of the subject and the alternative danger of making loose and inaccurate statements or dealing in glittering generalities too vague to carry conviction. As it is, the writer is here trying to give directly to the general public the results of years of special research in correlating the data from many scattered departments of science,--results that most scientists would feel obliged to reserve for the select few of some learned society, to be published subsequently in the Reports of its "Transactions," and to find their way after years of delay into the main currents of human thought. But these dilatory methods of professional pedantry, miscalled "ethics," shall not longer be allowed to delay the publication of highly important principles which the public are entitled to know at once, and to know at first hand. Then, too, it is more than doubtful if any purely academic body could be found willing to become responsible for giving to the world conclusions so contrary to the vogue of the present day.

That these brief chapters may clear up the doubts of some, and encourage the faith of many, is the object of their publication in this non-professional form.

G. McC. P.

Contents

I

MATTER AND ITS ORIGIN

I

When we were told by a prominent scientist just the other day that "electricity is now known to be molecular in structure," it almost took our breath away. And when we were informed that certain well-known chemical elements had been detected in the very act of being changed over into other well-known elements, with the prospect of such a transformation of the elements being quite the normal thing throughout nature, the very earth seemed to be slipping away from under our feet. Some of the closely related discoveries, such as the fact that the X-rays show a spectrum susceptible of examination, were not so disconcerting in themselves; but the marvellous pictures of the structure of the atom elicited by these discoveries made many good people almost question whether our venerable experimenters had not been indulging in pipe dreams amid their laboratory work.

Do we, then, begin to understand the real composition of matter? Does it have component parts, in the materialistic sense; or is what we call matter only a mysterious manifestation of energy? And if the latter be our answer, can we hope to settle the problem objectively and so conclusively that it will stay settled? In short, do we, regarding these border-line subjects between metaphysics and natural science, know anything more than our fathers and our grandfathers?

It will be convenient to consider these problems under two heads: the composition of matter, and the origin of matter.

II

1. It was long ago recognized that matter must be composed of particles which are driven farther apart by heat and are brought closer together by cold, thus laying the foundation for the theory of the molecular composition of matter. But not until the time of Dalton, about a hundred years ago, was it proved that the molecule itself, the unit of physical change, is capable of definite division into atoms, the units of chemical change. This conception of the molecules and atoms as the ultimate units of which matter is composed maintained its place until the discovery of radioactivity and its associated phenomena, about 1896; since which time we have definitely ascertained that even the atoms are separable into still smaller units, and that possibly these units are all alike. On this last possibility, it would surely be a most amazing fact if such multitudinous "properties" of bodies could be produced merely by variations in the arrangements of these ultimate units into atoms, or in some other way which produces vast differences in properties by combinations of units that are nevertheless mere duplicates of one another.

As hydrogen is the lightest of the elements, it has been a favorite theory with scientists that the various elements are all composed of combinations of hydrogen atoms. But since many of the elements have atomic weights which cannot be made exact multiples of that of hydrogen, it has been felt that there must be some other smaller unit than the hydrogen atom; or else that these hydrogen atoms themselves change in weight when they combine to form other atoms. But mass seems to be the one unchangeable characteristic of matter; hence it was felt that any change of weight is almost unthinkable, and so a solution was sought in the direction of still further dividing the hydrogen atom, the smallest unit concerned in chemical change, as then understood. But now the facts and principles brought to light in connection with the studies of radioactivity have settled it that we actually do have a much smaller unit than the hydrogen atom, one of only about $1/1760$ its mass, in fact; and that this smallest of the small things of nature is none other than a particle of negative electricity, now called an electron.

That the atoms of all the elements must have a common unit of composition, that they behave as if composed of ultimate particles that may be regarded as duplicates of one another, has long been regarded as an inevitable conclusion from the Periodic Law of Mendeleef. This law says that the physical as well as the chemical properties of the various elements depend upon their atomic weights, or as it is stated in the language of mathematics, the properties of an element are functions of its atomic weight. This fact of the variation in the properties of elements in accord with their atomic weights has been even more strikingly illustrated by the behavior of discharges of electricity through rarified gases, as well as by the facts of radioactivity. To quote the words of Sir J.J. Thompson, "The transparency of bodies to Roentgen rays, to cathode rays, to the rays emitted by radioactive substances, the quality of the secondary radiation emitted by the different elements, are all determined by the atomic weight of the element."[1]

[Footnote 1; Encyclopedia Britannica, Vol. XVII, 891. Cambridge Edition.]

Just recently we have had opened up before us a still more intimate inner-circle view of the composition of matter. H.G.J. Moseley, a young man only twenty-six years of age, at an English university, devised a method of examining the spectra of the various elements by means of the X-rays. He found in this way that the principal lines of these various spectra are connected by a remarkably simple arithmetical relationship; for when the elements are arranged in the order of their atomic weights, they show a graded advance from one to another equal to successive additions of the same electrical unit charge, thus indicating a real gamut of the elements that we can run up by adding or run down by subtracting the same unit of electrical charge. It is pitiable to have to record that next year this scientific genius was killed in the ill-fated Gallipoli expedition against Turkey.

Thus in many fairly independent ways we are brought around to this same idea of a common structure underlying all the many seeming diversities manifested by what we call matter.

The phenomena of radioactivity were discovered accidentally in 1896 by the French chemist Becquerel. Many investigators immediately began working along this promising line, and two years later Madam Curie, in association with others, discovered the new element radium. Soon it was discovered that radium and several other substances are continually giving off radiations at an enormous rate, that no change of chemical combination, no physical change of condition appears to have the slightest effect in slowing or increasing this discharge of emanations, while no scientific apparatus yet devised can detect any change in the substances left behind either in respect to weight or any other properties as the result of these enormous losses of energy. Accordingly some people not unnaturally were ready to draw the conclusion that those most firmly established laws of physics and chemistry, the laws of the conservation of energy and of matter, were overthrown by this astonishing behavior of these newly discovered substances. However, only a few more years of study and investigation were necessary to prove that this last conclusion was wholly unwarranted; and to-day these laws of the conservation of energy and of matter are more firmly established than ever.

The thing that has gone by the board is the old idea of the atoms as the indivisible and irreducible minima of the material universe. For not only do all the radioactive substances give off particles of helium gas positively electrified, but _all bodies, no matter what their composition_, can by suitable treatment, such as exposing them to ultra-violet light, or raising them to incandescence, be made to give off electrons or negatively charged particles, and these electrons are always the same no matter from what kind of substance they come. In a somewhat similar way, we always get positively electrified particles of the mass of the hydrogen atom, or about 1,760 times the mass of the electron, whenever we send an electric charge through a gas at very low pressure, no matter what the kind of gas. Whether or not these positive units will yet prove susceptible of being split up into smaller particles comparable to the electrons, is merely a subject for conjecture. We have no proof that they will. At the present time what we call matter seems to be composed of these positive units and of the electrons which are about 1/1760 as great; and in the present state of our knowledge these facts suffice to explain all the properties of matter. Thus we can either say that electricity is composed of matter, or say that matter is composed of electricity; and human language at best is such a clumsy vehicle of thought that scientifically and philosophically the one statement is as correct and as reasonable as the other.

And probably we shall never be able to learn any more than this. We have arrived at a sort of box-within-a-box theory of the make-up of matter. By a very elaborate system of unpacking, or by some violent external force that makes the inside burst open, as it were, we seem to be able to make pieces fly off from the atoms, these pieces being then projected into space with enormous force and velocity. There are theories galore of the structure of the atom; but as Prof. E.P. Lewis has said, most of these theories are so impossible as to be absurd, or so speculative that "they suggest no experimental tests for their validity."[2] Just at present Rutherford's theory of the structure of the atom is quite popular. This postulates a nucleus composed of a group of positive units and electrons, with an excess of the positive charges equal to half the atomic weight, with an equal number of electrons circulating about this nucleus in rings. Bohr's theory, which is not very different from this, has perhaps even more friends, and it is supported by the remarkable discoveries of the lamented Moseley. But we must not take such theories too seriously. As Kayser has said, any true theory of the make-up of the atoms must assume an absolutely full and perfect knowledge of all electrical and optical processes, and is therefore beyond our dreams. Or as Professor Planck said in his Columbia lectures, we are not entitled to hope that we shall ever be able to represent truly through any physical formul?the internal structure of the atom.

[Footnote 2: _Nature_, April 5, 1917.]

III

2. We must now take up the second phase of our subject, the problem of the origin of matter.

Before we knew anything of radioactivity we could have dismissed such a subject briefly by quoting the law of the conservation of matter, which says that matter can neither be created nor destroyed by any means known to science. By our knowledge of radioactivity we can make our answer a little more learned, a little less abrupt, but none the less discouraging to the advocate of the development hypothesis. We can tell how the elements of high atomic weight, such as uranium and thorium, are constantly giving off particles and are thus by loss or decomposition being changed over into other elements, such as radium, niton, polonium and lead. But our new knowledge compels us ultimately to give the same answer as before, namely, that _we still do not know how matter ever could have originated_, except that "in the beginning" it was called into existence by the fiat of Him whom we Christians worship as our God, the Creator. Thus we reach the conception of the universe as that of a great clock gradually running down, which is certainly the antithesis of that picture so long held before us by the advocates of the development theory.

Uranium is a rather rare element, though known for over a hundred years, and has an atomic weight of 238.5. In decomposing it gives off first a helium atom, weight 4; and after this action has been repeated three times the substance left is radium, atomic weight about 226.4. Thus radium is simply uranium after it has lost three helium atoms. Radium in its disintegration gives off three kinds of particles, namely, helium atoms (positively electrified), [Greek: b]-rays or electrons, and [Greek: g]-rays, the latter being identical with the X-rays, and having penetrating power sufficient to carry them through six inches of lead or a foot of solid iron. The final stage in this process of disintegration is the ordinary element lead, in which condition the atoms seem to have reached relative stability. Whether or not our stock of lead, with our other common elements that are not radioactive, was originally produced by the disintegration of these other elements, is merely a matter of conjecture. We know nothing at all about it.

The length of time it takes for half the atoms of an element to change is called its "life" or period. The periods of most of the radioactive substances have been calculated, that of uranium being very long. The calculated period of radium is 2,500 years, while that of polonium is only 202 days, and that of niton 5.6 days. These unquestioned facts, together with the enormous amount of heat evolved by the disintegration of these substances (that from radium being about 250,000 times the heat evolved by the combustion of carbon), have thrown a great deal of doubt upon the older estimates of the age of the earth.

The discussion of the details of these theories would be unprofitable. But through the mists of all these conflicting theories and probabilities two facts of tremendous importance for our modern world emerge in clear relief, namely, that the grand law of the conservation of matter still holds true, and hence that the matter of our world must have had an origin at some time in the past wholly different in degree and different in kind from any process going on around us that we call a natural process. These elements of high atomic weight that break down into others of lower atomic weight may be so rare because they have been about all used up in this process. At any rate, so far from revealing the origin of matter as a process now going on, these phenomena are an objective demonstration that all matter is more or less unstable and liable under some unknown but ever-acting force to lose some portion of that fund of energy with which it seems to have been primarily endowed. Not the evolution of matter but the degeneration of matter is the plain and unescapable lesson to be drawn from these facts. The varieties of matter may change greatly, and one variety or one chemical element may be transformed into another. But this transformation is by loss and not by gain. It is degeneration and not upward evolution that is now opened up before our astonished eyes by this peep into the ultimate laboratories of nature; and he is surely a blind observer who cannot read in these facts the grand truth that all this substance called matter with which science deals in her manifold studies must at some time in the past, I care not when, have been called into existence in some manner no longer operative. The past eternity of matter, as well as its progressive development from the simple to the complex, seems manifestly out of consideration in view of the facts as we now know them. There is no ambiguity in the evidence. So far as modern science can throw light on the question, there must have been a real Creation of the materials of which our world is composed, a Creation wholly different both in kind and in degree from any process now going on.

IV

A supposed objection has been raised to this view, based on the vastness of the universe as we now know it. Whether or not the universe is really infinite in extent, it is certainly of an extent that is practically infinite, so far as our powers of observation or of reasoning are concerned. But this practically infinite universe is not a bit harder to account for than would be a definitely limited universe, say of the size of our solar system. If the spectroscope shows that the far distant parts of the universe contain many of the same elements as are found in our solar system, we need not be surprised, since all are alike the work of the same Creator. Nor would this fact that the universe seems to be composed of similar materials throughout tend in any way to prove that all these parts of the universe were brought into existence at the same time, nor yet that our solar system was refashioned out of some of the common stock of the universe already on hand, as the nebular hypothesis supposes. For all that we can tell to the contrary, it would seem probable that the materials of our solar system were called into existence expressly for the position they are now occupying; and this seems to be the plain import of the record in Genesis. Of one thing, however, we can be certain,--these materials must at some time have been called into existence by methods or ways that are no longer in operation around us. "In the beginning God created the heaven and the earth."

V

Some remarks are necessary here regarding the homogeneousness of matter, or the idea that the various elements are composed of primordial units which are themselves alike, mere duplicates of each other. If this should prove to be really the case, as seems to be quite likely in the light of the facts given above, would it not be a veritable triumph for materialism? By no means. On the contrary, I think I can show in a very few words not only that this homogeneousness of matter is the only rational view of the composition of the material universe, but also that it is the only view consistent with Christian Theism and with the doctrine of Creation.

The theory of the atoms with their inherent and unchangeable properties, which prevailed during the greater part of the nineteenth century, naturally led us to look upon these properties as inherent in the things themselves. This was indeed materialism. This view, however, constantly impelled us to find out the essential differences between the various kinds of atoms, so as to "account for" their varying behaviors. And no matter how far we push such inquiries, this materialistic attitude of mind will control us so long as we think we are dealing with substances which are intrinsically different. If the differences are innate or inherent in the things themselves, we must naturally endeavor to find out why and how they are different; and no matter how far we go along this road we are always headed in the direction of stark materialism. On the other hand, to say that the "properties" of the atoms are not inherent in themselves, but are imposed on them by an external ceaselessly acting power, the will of the Creator, would be in full accord with Biblical theism; and then we might naturally say that the ultimate particles of which matter is composed may well be regarded as alike and mere duplicates of one another. And this, as we have seen, is just what modern discoveries in radioactivity are teaching us regarding the make-up of the substances that we call matter.

But an objection at once arises. How can these primordial units of which matter is composed behave so differently, if they are really alike, mere duplicates of one another?

We may not as yet be able to tell just why and how; but we have in the cells of which all plants and animals are composed an analogy which is almost perfect, if not quite.

These component units of organic matter, the individual cells, as will be explained later, seem physically and even chemically mere duplicates of one another. They may not all be of the same size; but they are all composed of protoplasm, and the protoplasm of plants cannot be distinguished from that of animals by any physical or chemical tests known to modern science. The protoplasm in the brain of a bird is the same as that in its toes; and no metaphysical subtilties about heredity have ever explained why the one does a different work from the other. The plain fact is that different cells, composed of identical protoplasm and structurally alike, _act very differently_; and there is no scientific reason based on innate properties that gives us even a glimmer of a reason why. We have searched a long time along this road; but there is no prospect of finding an explanation; we are merely running up a _cul-de-sac_ with no view beyond. From the materialistic point of view, nobody knows why protoplasm acts as it does, least _of all, why some masses of protoplasm act one way, and exact duplicates act differently_. But if, on the other hand, we look beyond the facts and methods of physics and chemistry, and even beyond the most plausible theories of genetics, we can readily explain this remarkable action of the cells as the result of the will of an ever acting, omniscient, almighty God. Certainly nothing else is adequate to explain the behavior of living cells.

In a very similar way we must reason regarding the ultimate units of matter, call them what we will, electrons, corpuscles, or units of electricity. If these are mere duplicates of each other, as science now teaches, they not only indicate by this identity that they are "manufactured articles," as was long ago pointed out of the atoms and molecules, but they also indicate with all the force of a demonstration that nothing but an ever present omniscient Intelligence could keep these duplicates from always acting the same under similar external forces. If gold and carbon, iron and oxygen are at bottom composed of particles that are mere duplicates of each other, as seems to be the case, how can these elements and the six dozen or more others maintain their individuality throughout nature as we know they do, even in the far distant stars, except by the sleepless care of an Intelligence whose Word is as effective in one part of the universe as in another, and to whose Word these particles of matter can show no inertia and no disobedience, because they have no powers or properties except what He has imparted? This doctrine of the homogeneousness of matter is the antithesis of materialism. It is consistent only with the doctrine of an almighty and ever present God, and like many other facts which have been developed by modern scientific discoveries, it confirms the other primal doctrine of a literal Creation "in the beginning."

VI

The conclusion which our minds are forced to draw from the facts presented in this chapter is not doubtful, nor is it difficult to state. Matter is not now being brought into existence by any means that we call "natural." And yet the facts of radioactivity very positively forbid the past eternity of matter. Hence, the conclusion is syllogistic: matter must have originated at some time in the past by methods or means which are equivalent to a real Creation.

Thus far, at least, the record of Genesis is confirmed: "In the beginning God created."

II

THE ORIGIN OF ENERGY

I

What has been regarded by many as the greatest scientific triumph of modern times was worked out about the middle of the last century by James Prescott Joule and others, in determining that a certain amount of mechanical energy is exactly equivalent to a definite amount of heat. With this mechanical equivalent of heat all the various other forms of energy have also been correlated; until now we have the general law of the Conservation of Energy, which says that energy can be neither manufactured nor destroyed, but merely transformed and directed. And this magnificent law, like that of the conservation of matter, is strong evidence that there must have been a real Creation at some time in the long ago, different not merely in degree but in kind from anything known to modern science.

Joule worked out the mechanical equivalent of heat by means of his now famous experiment of churning water. He reasoned that if the heat produced by friction, etc., is really energy in another form, then the same amount of heat must always be generated by the expenditure of a given amount of motion or mechanical work. And this must be true, no matter whether this work is expended in overcoming the friction between wood on wood, iron on iron, or in any other conceivable way. Accordingly, he devised an experiment in which paddle wheels were made to rotate in a vessel of water by means of falling weights somewhat like the weights of a clock. The amount of work represented by the falling of the weights was easily calculated, and so was the amount of rise in temperature of the water caused by the friction of the water with the rotating paddle wheels. In various other ways he measured the amount of heat generated by a measured amount of work; and as the result of all his experiments (with very slight corrections made since by means of more exact apparatus), we now know that 778 foot pounds of work produce heat enough to raise one pound of water one degree Fahrenheit; or stated in the metric system, 427 kilogram meters of work will produce a calorie of heat.

Since these record-making experiments by Joule, the matter has been verified over and over again in all sorts of ways; and almost every kind of display of energy has been measured with more or less exactness. Even the amount of food oxidized in the human body is now known to be capable of correlation with the other forms of energy, though necessarily very minute exactness of measurement is scarcely attainable in this case. But no scientist of to-day doubts that all the physiological processes of animals or of plants conform exactly to the law of the conservation of energy that energy is neither created nor destroyed by any means known to science. In other words, the amount of energy in our world, if science can at all determine such a matter, seems to be _a fixed quantity_, gradually being dissipated into space, it is true, but momently replenished from the sun at exactly the same rate now as hundreds or thousands of years ago. And while this energy is in our world it is always capable of exact correlation in all of its multitudinous forms, and is transformable back and forth without increase and without loss.

On the discovery of the radioactive substances in 1896, some persons hastily concluded that the law of the conservation of energy was contradicted by the astonishing way in which these substances acted. But further and more accurate experiments have set this matter at rest, as indeed might have been expected; for the law of gravitation itself is not more immovably established in the make-up of the universe than this magnificent law that energy cannot be created by any means which we call natural.

In all ages there have been men who have spent their lives in the vain effort to invent a machine out of which work could constantly be obtained without the expenditure upon it of an equal amount of work. But the United States patent office has got so tired of receiving applications for patents based on this idea of perpetual motion that they have long since refused to issue any such patent where this principle is the manifest object; and I suppose the governments of other countries have taken a similar stand. And why? Because they know that energy cannot now be created by any device, no matter how ingenious; and they refuse to become a party to any scheme that seems to imply that this modern creation of energy is within the bounds of possibility.

Yet what is all this but a confirmation of the declaration long ago made that "the works were finished from the foundation of the world" (Heb. 4:3)? True, the energy we are constantly employing seems to come to us from the sun; but we must remember that the sun and its family of the solar system, including the earth, were all made at the same time, that they are bound together as parts of an indissoluble whole. Accordingly, no one can say that the total amount of energy called into existence at the creation of our solar system is being added to at the present time. At any rate, so far as modern science can judge of the matter, the total amount of energy available for our world _is a fixed quantity_; and its amount and the terms on which it was to be available for our use were fixed or finished "from the foundation of the world." While it is a very significant fact in this connection that with all the multiform speculations which have been made as to the physical source of the sun's heat, no explanation wholly satisfactory has yet been made as to how this energy coming to us from the sun is constantly replenished or maintained.

The desire to find a material cause for all phenomena is instinctive in the human mind, and has proved the chief impetus in a thousand discoveries. And yet, unless we are on our guard, it is liable to be a source of real error whenever we are dealing with the deeper problems of thought. For when we have pushed our way into the inner sanctuary of any department of nature, we almost invariably come upon a deep chasm that we can pass over only by building a bridge of words. Some of these verbal bridges have been decorated with very dignified names, such as "the luminiferous ether," "gravity," "chemical affinity"; and when we have shifted from the one side of the chasm to the other we impose upon the credulity of the public (and even ourselves) by giving out the impression that these words represent the real objective bridge on which we crossed.

In how many ways do we by our theories dodge the crucial problem of how energy is really transmitted, that is, how matter can act on distant matter across seemingly vacant space. Gravity, and indeed all the forms of the attractive forces, come under this head. True, we observe certain regularities in the way in which these phenomena occur, and the phenomenon at one place seems to be somehow dependent on some exercise of force at another place. And so we invent an ingenious theory, and fortify it all around with ponderous algebraic artillery for defense against all attack. And by persistent use of such theories we hypnotize ourselves into the belief that we are truly scientific in method, and are dealing with objective realities, and that these learned theories are something more than pretentious masks to hide our ignorance of real nature; when in reality these theories seem to be only a material screen to shield us from an embarrassing near view of the immediate action of God in all the various phenomena of the world; for not many find it a comfortable thought thus to live continuously beneath the great Taskmaster's eye.

The theory of the luminiferous ether as the medium of the transmission of light is one of these pretentious bridges of words. Our advancing knowledge of electro-magnetic phenomena may some day drive us back to a modified form of the corpuscular theory of light, and then we can throw this of the ether to the winds. In that case we would at least have a real material cause for the phenomena with which we deal. While the current theory of the ether has so many inconsistencies, and attempts to bridge over so many real chasms in our thinking that it seems truly astonishing to see it taught so long. By the theory of the ether the problems are not solved, they are merely postponed or evaded; for while solving one difficulty it creates a multitude of its own. How then are we better off than before without any such theory?

Being at liberty to invent any sort of qualities for their ether, scientists have tried to imagine such a substance as they think they need. The ether must be a kind of matter; but unlike any matter that we know of it cannot have weight, or else it would gravitate together here and there, thus becoming more abundant in some places than in others; whereas the need is for a material absolutely uniform throughout space, even throughout the interiors of solid bodies, such as the earth and the bodies upon the earth.

Another reason for supposing the ether to be a _plenum_, filling absolutely all space, is that it must be perfectly frictionless; and for this reason it cannot be composed of particles with spaces between them. It must be frictionless, for otherwise the planets would be retarded in their motions through space. The earth, for instance, is moving along its orbit at the rate of eighteen miles a second; and yet the ether does not pile up in front of it, nor is it made rarer in the wake of the earth. Moreover, during the thousands of years during which astronomers have been making observations absolutely no retardation has been detected in the motions of the earth or of any of the heavenly bodies, even to the smallest fraction of a second.

It is necessary to make the ether absolutely elastic and absolutely rigid. We are acquainted with many materials that are elastic, and with some that are comparatively rigid. But the elastic substances that we are acquainted with are not rigid, and the rigid substances are not elastic; and to assume such contradictory qualities in the ether transports us far beyond the bounds of experimental science.

These are but a few of the difficulties raised by the assumption of the ether as a real entity; but as there is no means of demonstrating its existence, except by arguing the necessity of having such a medium to transmit radiant energy, it follows that no multiplication of objections to the theory is likely to refute it in the minds of those who feel this necessity. Those who refuse to admit the possibility of "action at a distance," who insist on inventing a connecting material medium between every observed effect and some material object with which it seems to be in causal connection, will, I suppose, have to be allowed to exercise their ingenuity in any way to satisfy their minds, even though they may have to revise their theory with every fresh discovery in optics or radioactivity.

There are many other ingenious mental devices, like this of the ether, which seem to me only materialistic efforts to postpone or to dodge the real vital lessons to be read from natural phenomena,--efforts to push the real Cause back one step farther into the shadow,--a last desperate effort, in the face of the constantly accumulating evidence of modern knowledge that the great First Cause is far more intimately connected with life and motion than many are willing to believe. We have already mentioned gravity and the other attractive forces, such as cohesion and adhesion; but seemingly very few people have ever paused to consider how utterly inexplicable they still remain in any physical or materialistic sense.

It is easy to explain any form of a push in a physical way; but gravity is not a push but a pull. And how are we to explain the method by which a body can act where it is not, how explain in detail the way by which it can reach out and pull in toward itself another separated body, and exert this pull across the immeasurably wide fields of space? The law of inverse squares may tell us very accurately the manner in which the results are accomplished, for our Creator is a God of order. But there is no materialistic theory of the why of gravitation that is worth employing the time of sensible, truth-loving people. And we can rest assured that there never will be any such real "explanation," save that this is the way which the great Jehovah has ordained. Since such theories only explain the known in terms of the unknown, they can serve only as a sort of mental buffer or shield between us and the conception of the direct working of a personal God, whose word must always be as effective throughout the remotest corners of His universe as near at hand, for the very simple reason that matter has no "properties" which He has not imparted to it, and accordingly it can have no innate inertia or reluctance to act which God's word would need to overcome in order to induce it to act, even when this word operates across the wide fields of space. On this explanation these phenomena of "action at a distance" are at least intelligible; while to me, and I speak now as a scientist, they are intelligible in no other way.

III

There is another line of thought which has to do with living organisms, but which I shall beg leave to anticipate and bring in here at the close of this chapter, since it follows as a direct corollary from the law of the Conservation of Energy. Indeed, we might even term it the biological aspect of that law.

As we have seen, we can neither create energy nor destroy it; though we can _lose it_,--so far as this earth is concerned. The vast fund of energy that daily comes streaming to us from the sun is transmuted back and forth in a thousand ways, though little by little it is dissipated off into space, and we are dependent upon a fresh supply from the ever replenished fountain.

Just so, though in a somewhat idealistic sense, is it with what we may term vital energy. Cells, organisms, even whole races, are subject to degeneration and decay. They cannot acquire higher powers, though they may gradually lose what they already have; as Bateson has recently told us that whatever evolution there is must be by loss and not by gain. Water very easily runs down hill; but cannot go up hill in and of itself. Just so with the types of organic life. It was not merely an idle sneer of the witty Frenchman, that science has not yet explained how an ancestor can transmit what he has not got himself. He cannot always transmit all that he himself actually possesses of nature's gifts. Vitality becomes lowered, and the type degenerates. Weismann has emphasized this idea in his doctrine of "panmixia," or the withdrawal of selection, which always results in degeneration. Selection, artificial or natural, may serve to counteract this universal tendency of organic life, but only approximately. As Sir William Dawson says, "All things left to themselves tend to degenerate." Little by little the endowment of vitality bestowed upon our world at the beginning has, like radiant energy, been returned to God who gave it; but, unlike the case of radiant energy, the Creator has not established any regular source of vital supply from without, no elixir of life for organic nature in general. There is no longer within easy reach a tree of life from which we may pluck and eat and live forever. And as the individual grows old and dies, so do species and even whole tribes degenerate and become extinct.

"From scarp cliff and quarried stone She cries, 'A thousand types are gone.'"

The glorious flood of vitality, so prodigally lavished upon our world in the beginning, has been ebbing lower and lower; and the theory of organic nature steadily advancing from the lower to the higher is manifestly just as puerile as the old hope of creating energy by a perpetual-motion machine,--and a mistake of precisely the same nature. Both are contradicted by the magnificent law of the Conservation of Energy, which, as we have said, is only the scientific expression of the Scriptural statement that Creation is completed, so far as our world is concerned; though, as the "wages of sin," death has been decreed upon the individual, and degeneration more or less marked upon every organic type. The fossils of the past, as well as our own experience within the historic period, confirm the view already arrived at on other grounds that _Creation is a completed work and is not now going on_; and the universal testimony from organic nature is that degeneration and decay have marked the history of every living form. Just as the individual grows old and dies, so do species degenerate and become extinct.

III

LIFE ONLY FROM LIFE

"No biological generalization rests on a wider series of observations, or has been subjected to a more critical scrutiny, than that every living organism has come into existence from a living portion or portions of a pre-existing organism."[3]

"Was there anything so absurd as to believe that a number of atoms, by falling together of their own accord, could make a sprig of moss, a microbe, a living animal? ... It is utterly absurd.... Here scientific thought is compelled to accept the idea of creative power. Forty years ago I asked Liebig ... if he believed that the grass and flowers, which we saw around us, grew by mere mechanical force. He answered, 'No more than I could believe that a book of botany describing them could grow by mere chemical force.'"[4]

"Let them not imagine that any hocus-pocus of electricity or viscous fluids would make a living cell.... Nothing approaching to a cell of living creature has ever yet been made.... No artificial process whatever could make living matter out of dead."[5]

[Footnote 3: P.C. Mitchell, in Encyclopedia Britannica, Vol. III, p. 952.]

[Footnote 4: Lord Kelvin in the London _Times_, May 4, 1903.]

[Footnote 5: Lord Kelvin, to a class of Medical Students, October 28, 1904.]

I

Ever since Ren?Descartes, in his Holland laboratory, dissected the heads of great numbers of animals in order to discover the processes of imagination and memory, men have been seeking a physical or materialistic answer to such questions as, What is life? What is it to be alive? How shall we distinguish the living from the not-living?

No one of to-day, in the light of the correlation of vital processes with the general law of the conservation of energy, believes that life in plants and animals is a separate entity which may exist outside of and apart from matter. In a scientific sense, we only know life by its association with living matter, which in its simplest form is known as protoplasm. The latter has been termed the physical basis of life, and so far as we know every material living thing is composed wholly of protoplasm and of the structures which it has built up.

This grayish, viscid, slimy, semi-transparent, semi-fluid substance, similar to the white of an egg, is the most puzzling, the most wonderful material with which science has to deal. Chemically it is composed of various proteids, fats, carbohydrates, etc., and these in turn of but very few elements, all of which are common, and none of which are peculiar to protoplasm itself. And yet its essential properties, its mechanical as well as its chemical make-up, have baffled the resources of our wisest men with all their retorts and microscopes and other instruments of precision.

Protoplasm is essentially uniform and similar in appearance and properties wherever found, whether in the tissues of the human body, in a blade of grass, or in the green slime of a stagnant pool. And yet probably no two samples of protoplasm are ever exactly similar in all respects, though we may never be able to detect their precise differences. These differences are due to the fact that the stuff is _alive_, and within it are constantly going on those changes accompanying metabolism, or the building up and tearing down processes that always accompany life. All separate masses of protoplasm, such as the one-celled amoeba or the individual cells of our own bodies, are constantly taking in food and as constantly throwing off wastes. Hence, in the very nature of things, it is impossible to find any mass of protoplasm absolutely pure. And a further and impassable barrier to chemical analysis, or indeed to any adequate scientific examination, lies in the fact that we can never deal with protoplasm exactly as it is, since no analysis can be performed upon it without destroying its life. And yet even dead protoplasm, and especially its most characteristic constituent, _proteid_, has been found the most difficult material in the world to analyze, and nobody as yet pretends to know its exact chemical make-up.

The constant effort of natural science to press back the boundaries of the unknown is very liable to obscure some of the things most essential to any system of clear thinking regarding these matters. We are so prone to think that if only our microscopes were a little stronger, if only we could devise more effective methods of staining or of chemical analysis or chemical synthesis, we might really find out what life is, or what matter itself is; in short, that we might be able to solve in a scientific way the old, old riddle of existence. But already we have about reached the limits of the powers of the microscope; and even if we could devise a way of seeing the ultimate structures of which protoplasm is composed, how would we be any better off? Would we not have to attribute to each constituent of this living substance the properties which we now attribute to the whole?--that is, the properties which we attribute to masses of protoplasmic units, such as plants, or birds, or human beings?

We look at ourselves and we feel sure that we have a separate and real existence, that we are rationally conscious and are endowed with choice and free will. We can say almost as much for an intelligent bird or dog. But we hesitate to say how many of these powers or characteristics of free and independent personality can be assigned to the unicellular organisms, such as the amoeba or the corpuscles of our blood. These one-celled creatures are also alive, are just as truly alive as are those composed of many cells. Even the corpuscles of which our bodies are composed move, and eat, and grow, and seem really endowed with intelligence like the higher forms of life. Suppose we could go further than is now possible and could lay bare the ultimate make-up of the chromatin of these one-celled creatures, would we even then be able to prove that life with all its properties is inherent in these material components of the cells? In other words, would we really solve anything after all? Or would we not rather be compelled to acknowledge that the simplest, the most truly rational view of the question is that in living matter we have merely a special manifestation of the presence and the direct action of the God of nature which we cannot so readily recognize in not-living matter? This, it seems to me, is all that we really know, and all that we are likely ever to know.

When we examine carefully the differences between the living and the not-living, we see that the chief difference between them is in their origin. The matter of growth is not a real distinction; for crystals grow on the outside, while inorganic liquids grow by intussusception, as when a soluble substance is added to them, in very much the same way as an animal grows by the ingestion of food. Even movement is hardly an absolute distinction between the living and the not-living; for no movement can be detected in quiescent seeds, which may lie dormant for thousands of years; and on the other hand inorganic foams when brought into contact with liquids of different composition display movements that very closely simulate those of the living matter. Lastly, irritability, though so notably characteristic of living matter, is scarcely peculiar to it, for many inorganic substances seem almost as definitely responsive to external stimulation. But in the matter of their origin there is a real and a most fundamental difference. All living substance arises only from other substance already living. It cannot arise from the not-living; or at least it never has done so since the beginning of scientific observation, though on this point have been concentrated the learning and the laboratory technique of thousands of chemists and microscopists.

It may not be out of place to quote here from one of the classics dealing with this subject,--words that are just as true to-day as when first written nearly half a century ago:

"Let us place vividly in our imagination the picture of the two great kingdoms of nature,--the inorganic and the organic,--as these now stand in the light of the Law of Biogenesis. What essentially is involved in saying that there is no spontaneous generation of life? It is meant that the passage from the mineral world to the plant or animal world is hermetically sealed on the mineral side. This inorganic world is staked off from the living world by barriers that have never yet been crossed from within. No change of substance, no modification of environment, no chemistry, no electricity, nor any form of energy, nor any evolution, can endow a single atom of the mineral world with the attribute of life. Only by the bending down into this dead world of some living form can these dead atoms be gifted with the properties of vitality; without this preliminary contact with life they remain fixed in the inorganic sphere forever.

"It is a very mysterious law which guards in this way the portals of the living world. And if there is one thing in nature more worth pondering for its strangeness, it is the spectacle of this vast helpless world of the dead cut off from the living by the Law of Biogenesis, and denied forever the possibility of resurrection within itself. The physical laws may explain the inorganic world; the biological laws may account for the development of the organic. But of the point where they meet,--of that strange border-land between the dead and the living,--science is silent. It is as if God had placed everything in earth and heaven in the hands of nature, but had reserved a point at the genesis of life for His direct appearing."[6]

[Footnote 6: Henry Drummond, "Natural Law in the Spiritual World,"

Chapter I

It would be superfluous to emphasize further this great outstanding fact that the not-living cannot become the living by any of the processes which we call natural; and it would be presumptuous to attempt to emulate these eloquent words by seeking to emphasize the completeness with which this great Law of Biogenesis confirms the truth of a real Creation; for the supreme grandeur and importance of this law could be only obscured by so doing.

II

Perhaps some of the most impressive lessons on this subject will be found in connection with the history of the discovery of this great Law of Biogenesis, which says that life can come only from life. For by studying the history of the way in which this great Law has been established, we cannot fail to be impressed with the thought that back of all the complex array of living forms in our modern world which go on perpetuating themselves in orderly ways according to natural law, they could have originated only by a direct and real Creation, essentially and radically different from any processes now going on.

The wisest of the ancients in Greece and Rome knew nothing of this great law as we now know it. Aristotle, the embodiment of all that the ancient world knew of natural science, expressly taught that the lower forms of animals, such as fleas and worms, even mice and frogs, sprang up spontaneously from the moist earth. "All dry bodies," he declared, "which become damp, and all damp bodies which are dried, engender animal life." According to Vergil, bees are produced from the putrifying entrails of a young bull. Such were the teachings of all the Greeks and Romans, even of the scientists of the post-Reformation period, some of whom had accumulated a very considerable stock of knowledge concerning plants and animals.

And similar absurdities continued to be taught until comparatively modern times. Van Helmont, a celebrated alchemist physician who flourished during the brilliant reign of Louis XIV, wrote: "The smells which arise from the bottom of morasses produce frogs, slugs, leeches, grasses, and other things." As a recipe for producing a pot of mice offhand, he says that the only thing necessary is partly to fill a vessel with corn and plug up the mouth of the vessel with an old dirty shirt. In about twenty-one days, the ferment arising from the dirty shirt reacting with the odor from the corn will effect the transmutation of the wheat into mice. The doctor solemnly assures us that he himself had witnessed this wonderful fact, and continues, "The mice are born full-grown; there are both males and females. To reproduce the species it suffices to pair them."

"Scoop out a hole in a brick," he says further, "put into it some sweet basil, crushed, lay a second brick upon the first so that the hole may be completely covered. Expose the two bricks to the sun, and at the end of a few days the smell of the sweet basil, acting as a ferment, will change the herb into real scorpions."[7]

[Footnote 7: "Louis Pasteur, His Life and Labors," p. 89.]

Sir Thomas Browne, the famous author of "Religio Medici," had expressed a doubt as to whether mice may be bred by putrifaction; but another scientist, Alexander Ross, disposed of this suggestion by the following line of argument which was supposed to be conclusive as a _reductio ad absurdum_:

"So may he (Sir Thomas Browne) doubt whether in cheese and timber worms are generated; or if beetles and wasps in cows' dung; or if butterflies, locusts, grasshoppers, shell-fish, snails, eels, and such like, be procreated of putrid matter, which is apt to receive the form of that creature to which it is by formative power disposed. To question this is to question reason, sense and experience. If he doubts this let him go to Egypt, and there he will find the fields swarming with mice, begot of the mud of Nylus, to the great calamity of the in-habitants."[8]

[Footnote 8: Encyclopedia Britannica, Vol. I, p. 64.]

When we remember that such nonsense constituted the wisdom of the scientific world only about two centuries ago, we begin to realize the fact that the doctrine of Biogenesis is indeed a very modern doctrine. But it may be well to ask in passing, How could the people of former ages understand or appreciate the great truth of Creation as we moderns are able to do?

The first important step toward the refutation of this old pagan doctrine of spontaneous generation was made by the Italian, Redi, in 1668. He noticed that flies are always present around decomposing meat before the appearance of maggots, and he devised an experiment to keep the flies away from actual contact with the meat. The meat putrified as usual, but did not breed maggots; while the same kind of meat exposed in open jars swarmed with them. He next placed some meat in a jar with some wire gauze over the top. The flies were attracted by the smell of the meat as usual, but could not reach the meat. Instead they laid their eggs upon the gauze, where they hatched in due time, while no maggots were generated in the meat. Thus from this time onward it became gradually understood that, at least in the case of all the larger and higher forms of life, Harvey's dictum, as announced some years previously, was true, and that life comes only from life.

But the invention of the microscope opened the way for a renewal of the controversy regarding the origin of life. Bacteria were discovered in 1683; and it was soon observed that no precautions with screens or other stoppers could prevent bacteria and other low organisms from breeding in myriads in every kind of organic matter. Here apparently was an entirely new foundation for the doctrine of spontaneous generation. It was freely admitted that all the higher forms of life arise only by process of natural generation from others of their own kind; but did not these microscopic organisms prove that there was "a perpetual abiogenetic fount by which the first steps in the evolution of living organisms continued to arise, under suitable conditions, from inorganic matter"?[9]

[Footnote 9: Encyclopedia Britannica, Vol. I, p. 64.]

The famous "barnacle-geese" ought not to be omitted from any sketch of the vicissitudes of this doctrine of Biogenesis. An elaborate illustrated account covering their alleged natural history was printed in one of the early volumes of the Royal Society of London. Buds of a particular tree growing near the sea were described as producing barnacles, and these falling into the water were alleged to be transmuted into geese. Nor should we omit mention of Huxley's _Bathybius Haeckelii_, a slimy substance supposed to exist in great masses in the depths of the ocean and to consist of undifferentiated protoplasm, the exhaustless fountain from which all other forms of life had been derived. Not long after Huxley had given it a formal scientific name in 1868, it was discovered to be merely a precipitate of gypsum thrown down from sea water by alcohol, and thus a product of clumsy manipulation in the laboratory, instead of a natural product of the deep sea. The disappointment of those opposing biogenesis was severe; but the lesson is still of value to the world to-day.

The masterly work of Tyndall and Louis Pasteur in doing for the bacteria and protozoa what Redi had done for the larger organisms, is too much a matter of modern contemporary history to need recital here. Upon this great truth of life only from life is based all the recent advances in the treatment and prevention of germ diseases and all the triumphs of modern surgery. The housewife puts up canned fruit with the utmost confidence because she believes in this great Law of Biogenesis. It is because we all believe in it that we use antiseptics and fumigators and fly screens.

III

But what are the lessons to be learned from this great fact, and what bearing has this fact on the old Bible doctrine of a literal Creation?

Life comes now only from preexisting life. But at some time there was no life on the globe. It does not take any great exercise of "philosophic faith," as Huxley suggested, "to look beyond the abyss of geologically recorded time" and recognize that at this beginning of things there must have taken place a most wonderful event, essentially and radically different from anything now going on, namely, the beginning of organic life. But would not this be a real Creation in the old-fashioned sense of this term? We cannot avoid this conclusion; nor is there anything in either science or philosophy to indicate that this creation of the living from the not-living was confined to one mere speck of protoplasm. It is absolutely certain that it required a real Creation to produce life from the not-living at all; and it is just as reasonable that this exercise of creative power may have taken place _in all parts of the earth at the same general time_, as the Bible teaches. For if a Being saw fit to create life at all, why should He stop with one or two bits of protoplasmic units? An architect who can make his own bricks and other building material, can surely build what he desires out of these materials. Common sense tells us that, if the Creator really created life in the beginning, He did not stop with a few specks of protoplasm here and there over the earth. The ability to create life from the not-living implies the ability to make full-grown trees or birds or beasts in twenty-four hours, instead of waiting for months or years, as is usual at the present time.

As we have already found regarding matter and energy, so of life. The record in Genesis is confirmed, for modern science compels us to believe in Creation as the only possible origin of life,--a Creation entirely different from anything now going on, and one that can never be made to fit into any scheme of uniformitarian evolution.

IV

THE CELL AND THE LESSONS IT TEACHES

I

With his usual vigor and expressiveness Henry Drummond has given us a picture of the remarkable fact that the cells of all plants and animals are strikingly alike, especially the single cells from which all originate. It is easy for any one to distinguish between an oak, a palm tree, and a lichen, while a botanist will have elaborate scientific distinctions which he can discern between them. "But if the first young germs of these three plants are placed before him," says Drummond, and the botanist is called upon to define the difference, "he finds it impossible. He cannot even say which is which. Examined under the highest powers of the microscope, they yield no clue. Analyzed by the chemist, with all the appliances of his laboratory, they keep their secret.

"The same experiment can be tried with the embryos of animals. Take the ovule of the worm, the eagle, the elephant, and of man himself. Let the most skilled observer apply the most searching tests to distinguish the one from the other, and he will fail.

"But there is something more surprising still. Compare next the two sets of germs, the vegetable and the animal, and there is no shade of difference. Oak and palm, worm and man, all start in life together. No matter into what strangely different forms they may afterwards develop, no matter whether they are to live on sea or land, creep or fly, swim or walk, think or vegetate,--in the embryo, as it first meets the eye of science, they are indistinguishable. The apple which fell in Newton's garden, Newton's dog Diamond, and Newton himself, began life at the same point."[10]

In these remarks, of course, Drummond is dealing with the unicellular primal form, "as it first meets the eye of science"; and while certain slight peculiarities (such as the constant number of chromosomes) have been detected as characteristic of the cells of certain forms, yet for all practical purposes these words of Drummond are just as true to-day as when first written. Possibly it is because of a failure in our technique or from a lack of power in our microscopes that these wonderful protoplasmic units from which all living things originate seem identical. But it is equally possible that they are really identical in structure and in chemical composition, and that only the ever present watchcare of the great Author of nature directs the one to develop in a certain manner, "after its kind," and another in still another manner, "after its kind." At any rate, the protoplasm of which they are all alike composed is identical wherever found, so far as any scientific tests have yet been able to determine.

[Footnote 10: "Natural Law,"

Chapter X

.]

II

There are many varieties of single cells known to science which maintain an independent individual existence. Among the unicellular plants are the bacteria, while the unicellular animals are known as the protozoa. And although perhaps I ought to apologize to the reader for seeming to anticipate here a part of the discussion of the problem of "species," yet it seems necessary to say a few words here regarding the "persistence" of these unicellular forms.

Among the diseases which have been proved to be due to protozoa are malaria, amoebic dysentery, and syphilis; while among the much larger number which are due to bacteria, bacilli, or other vegetable parasites, are cholera, typhoid fever, the plague, pneumonia, diphtheria, tuberculosis, and leprosy.

One of the difficulties attending the study of "species" among the higher forms of plants and animals has always been the length of time required to obtain any large number of generations on which to make observations. In the case of such plants as peas, wheat, corn, or indeed almost any form of plant life, it is only with difficulty that more than one generation a year can be obtained; and when two or more generations a year are produced, they are produced under more or less unnatural conditions. So that it takes almost a lifetime carefully to test and record in a thoroughly scientific way the results of any extensive experiments regarding variation and heredity.

In the case of mice or rats or rabbits or guinea pigs, many more generations can be obtained in a few years; but in the case of the larger kinds of animals the time taken for development to maturity and for gestation is often much prolonged; and scientific observation of an exact character has been in vogue for so short a time that there has always been the chance for advocates of evolution to take refuge under the plea that, if we only had longer and more carefully conducted observations, we could really see species in the making, one form becoming transformed into a distinct form, or perhaps giving rise to another and distinct form as an offshoot.

But in the case of the bacteria and protozoa, we can have a new generation every hour or so, sometimes every half hour. True, these forms of minute life have been under observation for only a few years; but their effects have in many cases been observed for almost the entire length of human history. No physician would tolerate the suggestion that the bacillus of cholera can produce the symptoms of diphtheria, or the tubercle bacillus produce the symptoms of leprosy. Nor will any scientist deny that such diseases as the plague, tuberculosis, or diphtheria are identical with diseases which ravaged Rome or Greece or Egypt thousands of years ago. And as the symptoms of these modern diseases are similar to those recorded by acute observers in Greece or Egypt two thousand years or more ago, we must conclude that the organisms causing these symptoms are doubtless identical. Similar remarks might be made regarding fermentation and other forms of decay.

In the case of a form of bacteria which reaches maturity and redivides in half an hour, the number of individual forms existing at the end of two days would need about twenty-eight figures to represent it. Doubtless these forms never multiply at this rate uninterruptedly for any great length of time, or else they would occupy the whole world to the exclusion of every other form of life. And doubtless instances arise where the period of growth to maturity and division is prolonged to several times the half-hour period mentioned above. But in any case, as we contemplate the length of time during which such well marked diseases as diphtheria, leprosy, or the plague have been known, we must acknowledge that these unicellular forms seem to breed true during a most astonishingly long period. How can we deny that this "persistence" of these unicellular forms constitutes a very strong argument in favor of the "fixity" of these forms?

III

But we must proceed to examine the behavior of the various kinds of cells of which the various multicellular organisms are composed.

Plants were known to be composed of cells, and their cells were studied and described some years before it was understood that animals also are composed of cells as units. Even then, however, the first propounders of the cell theory (Schleiden and Schwann) had no clear or accurate idea of the origin of cells, or of their essential characters and structure. As to origin, they supposed that cells arose by a sort of crystallization from a mother liquor; and as to structure, they looked upon the cell-wall as the really important part, the fluid contents being quite subordinate. Hugo von Mohl (1846) applied to the fluid contents of the cell the term "protoplasm," and Max Schultze (1861) showed that this protoplasm is really identical in all organisms, plants and animals, also that the cell-wall is frequently absent in many animal tissues and in many unicellular forms, indicating that the protoplasm is the really important substance. By this time also it had become known that cells never arise _de novo_, as had been supposed by the earlier investigators, but that cells arise only by division of preexisting cells; or as Rudolf Virchow (1858) expressed it, "_omnis cellula e cellul[=a]._"

It was, however, many years before the details of the growth and reproduction of the cells (cell-division) became well understood. Not until the last quarter of the nineteenth century was it settled that the nucleus of the cell is also a supremely important part; but finally in 1882 Flemming was able to extend Virchow's aphorism to the nucleus also: omnis nucleus e nucleo.

Since these discoveries our knowledge of the methods of cell-division has much increased; and in the light of our modern knowledge of these matters there is nothing in all nature more marvellous than the regular orderly way in which cells reproduce themselves according to fixed laws. Certain cells in the developing embryo, for example, are early set apart for a particular function or for building certain structures, and thereafter are never diverted from this duty so as to do a different work or produce a different kind of structure. In the young embryo certain structures arise at certain predestined times in particular places, and only there and out of these cells alone. As to why it should be so, we cannot tell, save as the result of deliberate design and as an expression of the order-loving mind of the God of nature. In the words of one of the greatest of modern authorities, "We still do not know why a certain cell becomes a gland-cell, another a gangleon-cell; why one cell gives rise to smooth muscle-fiber, while a neighbor forms voluntary muscle.... It is daily becoming more apparent that epigenesis with the three layers of the germ furnishes no explanation of developmental phenomena."[11]

[Footnote 11: _Nature,_ May 23, 1901.]

In accordance with the general principle of a division of labor, certain cells become early set apart to particular functions, and in accordance with the varying demands of these functions the developing cells may become greatly changed in form and in vital characteristics. That is, one cell specializes, let us say, in secretion, another in contractility, another in receiving and carrying stimuli, etc. In this way we will have the gland-cell, the muscle-cell, and the nerve-cell, each cell destined to produce one of these organs developing others "after its kind," the result being that it is soon surrounded with numerous companions doing a similar work, making up in this way a particular tissue or organ--gland, muscle, or nerve--which in the aggregate has for its function the work of the particular cells composing it.

But the important thing for us to remember in this connection is that when cells once become thus differentiated off and dedicated to any particular function, they can never grow or develop into any distinctly different type of cell with other and different functions. It is true that through pathologic degeneration the form and even the function of cells may become greatly changed; but never does it amount to a complete metamorphosis or complete transformation into another distinctly different type.

This is a very important principle, and it contains so many lessons for us bearing on the philosophy of life in general that it may be allowable to establish this fact by several somewhat lengthy quotations from standard authorities.

The first will be from one of the highest authorities on embryology, Charles Sedgwick Minot, of Harvard:

"In accordance with this law [of differentiation] we encounter no instances, _either in normal or pathological development_, of the transformation of a cell of one kind of tissue into a cell of another kind of tissue; and further we encounter no instances of a differentiated cell being transformed back into an undifferentiated cell of the embryonic type with varied potentialities."[12]

Again, we have the following from one of the foremost pathologists, as to the strict and rather narrow limits of even pathologic change:

"Epithelium and gland cells ... never become converted into bone or cartilage, or vice versa; while, again, it may be laid down that among epiblastic and hypoblastic tissues, on the one hand, and mesoblastic tissues on the other, there is no new development or metaplasia of the most highly specialized tissues from less specialized tissues; a simple epithelium cannot in the vertebrate give rise to more complex glandular tissue, or to nerve cells; in regeneration of epithelium there is no new formation of hair roots or cutaneous glands. The cells of white fibrous connective tissue have not been seen to form striated or even non-striated muscle."[13]

[Footnote 12: _Science_, March 29, 1901, p. 490.]

[Footnote 13: J.G. Adami, "Principles of Pathology," pp. 641-642.]

As implied by these quotations, a constant and progressive differentiation of cells prevails in the developing embryo; and when complete, certain groups of cells act as specialists in doing only certain kinds of work for the body. These cells maintain their specific characters in a very remarkable degree under normal conditions. Under various abnormal conditions, however, these cells may become modified as to functions, so that cells or tissues of one type may assume more or less completely the characters of another type. "But," as a very high authority declares, "the limitations in this change in type are strictly drawn, so that one type can assume only the characters of another which is closely related to it. This change of one form of closely related tissue into another is called metaplasia....

"When differentiation has advanced so that such distinct types of tissue have been formed as connective tissue, epithelium, muscle, nerve, _these do not again merge through metaplasia. There is no evidence that mesoblastic tissues can be converted into those of the epiblastic or hypoblastic type, or vice versa_."[14]

[Footnote 14: Delafield and Prudden, "Text-Book of Pathology," pp. 62, 63.]

This modification of function among the cells which sometimes goes on in the developing embryo, or under pathologic conditions, is very closely analogous to the variation which goes on among species of animals and plants. But, as we shall see later, there is a well marked limit to this variation among species, just as we see there is in the variations among the cells. Practically the same general laws hold good in each case.

If cells did not maintain their ancestral characters in a very remarkable way, what would be the use of grafting a good kind of fruit onto a stock of poorer quality? The very permanency of the grafts thus produced is proof of the persistency with which cells reproduce only "after their kind."

IV

How can we fail to see the bearings of these facts on the doctrine of the transformation of species among ordinary plants and animals, which are merely isolated and self-contained groups of cells? Do not these facts constitute strong presumptive evidence that among animals and plants, though there may be variation in plenty within certain limits, perhaps within even much wider limits than used to be thought possible, yet among these distinct organisms, little and big, new forms develop only after their ancestral type, in full accord with the record given in the first chapter of the Bible?

But we are now prepared to examine in more detail the facts as now known to modern science regarding "species" of plants and animals.

V

WHAT IS A "SPECIES"?

I

We have seen that there is no way to account for the origin of matter, of energy, or of life, except by postulating a real Creation.

We have seen that cells continue to maintain their identity, and reproduce only "after their kind."

We must now deal with the higher forms of cell aggregates, which we call plants and animals. It has long been held that these at least are mutable, that one kind of plant or of animal may in the course of ages be transformed into a distinctly different type; and of late years there has accumulated a very voluminous literature dealing with the various intricacies of this problem of the origin of species. How can we deal with such a large subject in a brief way? It seems best to confine our attention in this chapter to an attempt to answer the question, What is a species? and are "species" natural groups clearly delimited by nature?

II

The term "species" was at first used very loosely by scientific writers. It meant very little more than our vague word kind does at the present time. Not until the time of Linn鎢s (1707-1778) did the term acquire a definite and precise meaning. The aphorism of the great botanist, "_species tot sunt divers?quot divers?form?ab initio sunt create"--"just so many species are to be reckoned as there were forms created in the beginning,"--was at least an attempt to use the term in a well-defined sense. Of course, this definition assumed the "fixity" of species; but with the wide prevalence of the views of Darwin and his followers the term "species" has fallen into disrepute, and is now regarded by many as only an artificial rank in classification corresponding to no objective reality in the natural world. Some writers, as Lankester, have found so much fault with the term as to urge its complete abandonment in scientific literature. This is logical enough from the standpoint of Darwinism; for if the latter be true there ought indeed to be such a swamping of every incipient "species" as to make one kind blend with others all around it in the classification series.

But since the term has by no means been discarded, we must endeavor to determine the sense in which it continues to be used in good scientific literature.

"A species," says Huxley, "is the smallest group to which distinct and invariable characters can be assigned." The Standard Dictionary says that the term is used for "a classificatory group of animals or plants subordinate to a genus, and having members that differ among themselves only in minor details of proportion and color, and are capable of fertile interbreeding indefinitely."

The latter authority also adds:

"In the kingdoms of organic nature species is founded on identity of form and structure, and specifically characterized by the power of the individuals to produce beings like themselves, who are in turn productive."

To put the matter still more definitely before the reader, we quote the following from a well-known scientist whose writings on the subject of evolution have had a wide circulation:

"There are two bases on which species may be founded. Species may be based on _form_, morphological species; or they may be based on _reproductive functions_, physiological species. By the one method a certain amount of difference of form, structure, and habit, constitutes species; according to the other, if the two kinds breed freely with each other and the offspring is indefinitely fertile, the kinds are called varieties, but if they do not they are called species."[15]

This author adds that this physiological test, as to whether or not the kinds are cross fertile, "is regarded as a most important test of true species, as contrasted with varieties or races."

[Footnote 15: Joseph Le Conte, "Evolution and Religious Thought," p. 233.]

When we look at the matter in this light, it is very evident that there are multitudes of long recognized specific distinctions that ought to be discarded. For instance, there are some twenty odd "species" of wild pigs scattered over the Old World, which Flower and Lydekker assure us would probably "breed freely together."[16] The yak and the zebu of India, and the bison of America, would on this basis have to be surrendered, for it is well known that they will all breed freely with the common domestic cattle, as well as with one another. Perhaps all or nearly all of the dozen or more "species" of the genus Bos would thus be included together. All of the dogs, wolves, and others of the Canida might thus be considered as fundamentally a unit. The cats (Felid) are well known to breed freely together, Karl Hagabeck of Hamburg having crossed lions and tigers as well as others of the family. Practically all of the bears have been crossed repeatedly, and the progeny of these and other crosses are quite familiar sights at the London Zoological Gardens. Among the lower forms of life even more surprising results have been attained by Thomas Hunt Morgan and others.

[Footnote 16: "Mammals Living and Extinct," pp. 284-285.]

It would, however, be a very hasty conclusion to say on the basis of these facts that there are no natural limitations to groups of animals and plants. But we are entirely warranted in concluding from these facts that in very many cases, perhaps in most, our system of taxonomic classification of animals and plants has gone altogether too far, and that scientists have erected specific distinctions which are wholly uncalled for and which confuse and obscure the main issues of the species problem. Among the workers in botany and in every department of zoology there have always been the "splitters" and the "lumpers," as they are familiarly called; the former insisting on the most minute distinctions between their "species," thus multiplying them; the latter being more liberal and tending to diminish the number of species in any given group. For a generation or more in the recent past the "splitters" had things pretty much their own way; but of late there is a growing tendency to frown down the mania for creating new names. Even yet it is with the utmost reluctance that long established specific distinctions are surrendered, as is illustrated in the case of the mammoth, which is acknowledged by some of the very best authorities to be really indistinguishable from the modern Asiatic elephant. Several fossil bears were long listed in scientific books; but they are all acknowledged now to be identical with the modern grizzly, and as we have already intimated all the modern ones ought to be put together. These modern rationalizing methods have made but a slight impression on the vast complex of the fossil plants and animals, affecting the names of only a few of the larger and better known forms. In the realm of invertebrate paleontology, however, the "splitters" are still holding high carnival, in spite of the efforts of some very prominent scientists in the opposite direction. For paleontologists still follow the irrational course of inventing a new name, specific or even generic, for a form that happens to be found in a kind of rock widely separated as to "age" from the other beds where similar forms are accustomed to be found. As Angelo Heilprin expresses it, "It is practically certain that numerous forms of life, exhibiting no distinctive characters of their own, are constituted into distinct species for no other reason than that they occur in formations widely separated from those holding their nearest kin."[17]

As a result of these methods this same author declares: "It is by no means improbable that many of the older _genera_, now recognized as distinct by reason of our imperfect knowledge concerning their true relationships, have in reality representatives living in the modern seas."[18]

[Footnote 17: "Geographical and Geological Distribution of Animals," pp. 183, 184.]

[Footnote 18: Id., pp. 207, 208.]

But the situation is very little better when we come to deal with plants and animals of our modern world. Because, with the many thousands of students of natural science all over the world, each anxious to get into print as the discoverer of some new form, the systematists have a dead weight of names on their hands that by a rational and enlightened revision could doubtless be reduced to but a fraction of their present disheartening array. For as the result of the extensive breeding experiments now being carried on under the study of what is called Mendelism (a term that will be explained in the next chapter), it has been found that great numbers of the "species" of the systematists or classificationists will not stand the physiological test of breeding, that is, they are found to breed freely together according to the Mendelian Law. As William Bateson remarks:

"We may even be certain that numbers of excellent species recognized by entomologists or ornithologists, for example, would, if subjected to breeding tests, be immediately proved to be _analytical varieties_, differing from each other merely in the presence or absence of definite factors."[19]

The following from David Starr Jordan, the leading American authority on fishes, will serve to show how numerous have been the new names invented in recent years, all tending further to confuse and complicate the problem of what is a species:

"In our fresh-water fishes, each species on an average has been described as new from three to four times, on account of minor variations, real or supposed. In Europe, where the fishes have been studied longer and by more different men, upwards of six or eight nominal species have been described for each one that is now considered distinct."[20]

[Footnote 19: "Mendel's Principles of Heredity," p. 284, 1909.]

[Footnote 20: "Science Sketches," p. 99.]

And again:

"Thus the common Channel Catfish of our rivers has been described as a new species not less than _twenty-five times_, on account of differences real or imaginary, but comparatively trifling in value."[21]

[Footnote 21: "Science Sketches," p. 96.]

Perhaps the reader will tolerate another somewhat long quotation because of the light which it sheds on this whole problem.

"Some years ago we had a parasite of a very destructive aphid down in our books as Lysiphlebus tritici. In carrying out our investigations it became necessary to find out whether this parasite had more than a single host insect, and whether it could develop in more than one species of aphid. To this end, recently emerged males and females were allowed to pair, after which the female oviposited in several species of aphids. Both parents were then killed and preserved and all of their progeny not used in further experiments were also preserved, and thus entire broods or families were kept together. In this way females were reared out of one host species and allowed to oviposit in others, until, often after several hosts had been employed, it would be bred back into the species whence it first originated. In all cases the host was reared from the moment of birth, while with the parasite both parents and offspring were kept together.

"The result of this little fragment of work _was to send two genera and fourteen species to the cemetery_--you may call it Mt. Synonym Cemetery, if you choose--while the insect involved is now Aphidius testaceipes. The systematist who studies only dried corpses will soon be out of date."[22]

[Footnote 22: F.M. Webster, of the U.S. Dept. of Agriculture, in _Science_, April 12, 1912, p. 565.]

IV

Now all this is not given to intimate that there is no scientific justification for the term "species," but to make plain to my non-professional readers what every well-informed biologist already knows, namely, that at the present time the "species question" is still in a very unsatisfactory state. The facts given above would strongly suggest that there probably is indeed such a thing as a species, in the sense assigned by Linneus, who as we have seen wished to make it a designation covering all the descendants of each distinct kind originally created. But this original aim of Linneus is to-day not merely ignored but treated with lofty contempt; for according to the prevailing theories of evolution, all the manifold diversities of life in our modern world have come about gradually as the result of a slow development by natural process, and hence it would be vain beyond measure to attempt to determine the limits of a "species" in the sense understood by Linneus.

But we may conclude, from the facts presented above, that if there is such a naturally delimited group as a "species" in the Linn鎟n sense of the word, it by no means coincides with what now passes under this name, but might include many so-called species, often a whole genus, or even several.

With this in mind, we must pass on to consider the next step in our study, as to whether new "species" are now coming into being in our modern world under scientific observation, either natural or artificial.

VI

MENDELISM AND THE ORIGIN OF SPECIES

"Had Mendel's work come into the hands of Darwin, it is not too much to say that the history of the development of evolutionary philosophy would have been very different from that which we have witnessed."[23]

[Footnote 23: William Bateson, "Mendel's Principles of Heredity," p. 316.]

I

From the latter part of the eighteenth century, attempts were continually being made to explain the origin of all organic forms by some system of development or evolution. Buffon had dwelt on the modifications directly induced by the environment. Lamarck had made much use of this idea, claiming that such modifications were transmitted to posterity, and claiming the same for the structural changes produced by use and disuse. Lamarck's work did not become at all popular while he lived, chiefly through the overpowering influence of Baron Cuvier, who had an equally fantastic scheme of his own, which may well be termed a burlesque on Creation and in which an extreme fixity of "species" was a cardinal doctrine. Erasmus Darwin and Robert Chambers in England also tried to make a theory of evolution believable; though their efforts were but little more successful in gaining the ear of the world.

But to all that had gone before Charles Darwin and A.R. Wallace (1858) added the idea of "natural selection," or "the struggle for existence," to use the respective terms coined by each of these authors, as the chief means by which the effects of variation are accumulated and perpetuated so as to build up the modern complexities of the plant and animal kingdoms. Partly because it was a psychological moment, from the fact that the uniformitarian geology of Lyell with its graded advance of existences from age to age seemed absolutely to demand some evolutionary explanation; partly because artificial selection was a familiar idea of proved value in selective breeding, and "natural selection" seemed an exact parallel carried on by nature in the direction of continual improvement; but perhaps more largely because the abstract idea of "natural selection" involved so many intricate separate concepts that for nearly a generation scarcely two naturalists in the world could state the whole problem of the theory exactly alike;--on all these accounts the theory of natural selection, or of the "survival of the fittest," to use the phrase of Herbert Spencer, became in the latter decades of the nineteenth century well-nigh universal.

But about 1887 a faction or school arose who criticized the main idea of Darwin and Wallace and fell back on the Lamarckian factor of the transmission of acquired characters as really the essential cause of the process of evolution. Herbert Spencer, E.D. Cope and others did much to criticize natural selection as inadequate to do what was attributed to it, dwelling on the importance of the transmission of acquired characters. Spencer even went so far as to declare, "either there has been inheritance of acquired characters, or there has been no evolution." These Neo-Lamarckians argued that natural selection alone can neither explain the origin of varieties, nor the first steps in the slow advance toward "usefulness." An organ must be already useful before natural selection can take hold of it to improve it. Selection cannot make a thing useful to start with, but only (possibly) make more useful what already exists. Until the newly formed buds of developing limbs or organs became decidedly "useful" to the individual or the species, would they not be in the way, merely so many hindrances, to be removed by natural selection instead of being preserved and improved? But, in this view of the matter, they argued, what single organ of any species would there be that must not thus have appeared long before it was wanted?

Or to use the pungent words quoted with approval by Hugo de Vries at the end of his "Species and Varieties" (pp. 825, 826), "Natural selection may explain the survival of the fittest, but it cannot explain the arrival of the fittest."

This side of the argument is dwelt upon at some length by Alex. Graham Bell, as reported in a recent interview. He says:

"Natural selection does not and cannot produce new species or varieties or cause modifications of living organisms to come into existence. On the contrary, its sole function is to prevent evolution. In its action it is destructive merely,--not constructive,--causing death and extinction, not life and progression. Death cannot produce life; and though natural selection may produce the death of the unfit, it cannot produce the fit, far less evolve the fittest. It may permit the fit to survive by not killing them off, if they are already in existence; but it does not bring them into being, or produce improvement in them after they have once appeared."[24]

[Footnote 24: _World's Work_, December, 1913, p. 177.]

Opposing these Neo-Lamarckians were such prominent scientists as August Weismann, A.R. Wallace, E. Ray Lankester, who strenuously opposed the idea that "acquired characters," or more precisely _parental experience_, are ever transmissible. In the subsequent years the greatest variety of experimental tests have been applied to secure the hereditary transmission of any sort of such acquired characters, with uniformly negative results. One of the most elaborate of these experiments was conducted by a German botanist, who transplanted 2,500 different kinds of mountain plants to the lowlands, where he studied them for several years alongside their relatives, natives of these lowlands. He found that their mountain environment had made absolutely no permanent change in their structures or habits, which soon conformed exactly with those of their relatives which had lived in the lowland environment for centuries. Many similar efforts have been made to confirm this doctrine of the transmission of acquired characters; but their universal failure is like that of mechanics in trying to invent perpetual motion.

Thomas Hunt Morgan sums up the present situation in the following words: "To-day the theory has few followers among trained investigators, but it still has a popular vogue that is wide-spread and vociferous." And we may add that the extent of its spread is directly proportioned to the need felt for this doctrine as a support of the theory of evolution, while the vociferance of its advocates is inversely proportioned to the evidence in its support.

As a result of extensive modern experiments and discussion, biologists have grown very cautious, and are by no means so positive as they were twenty years ago in affirming just how species have come into existence. Echoes of this old controversy between the two leading schools of biologists are occasionally heard; but the enthusiasm with which they set out a half century ago to solve the riddle of plant and animal life has largely given way to a purpose to discard speculation and patiently to observe and record actual facts. For with natural selection discredited in the house of its friends, and Lamarckianism under grave suspicion from want of a single well authenticated example, it is hard to see what there is left of the biological doctrine that has so dominated scientific thought for a half century. If each of these opposed schools of scientists are right in _what they deny_, the whole theoretical foundation for the origin of new kinds of animals and plants is swept away,-- absolutely gone. For if an individual really cannot transmit what he has acquired in his lifetime, how can he transmit what he has not got himself, and what none of his ancestors ever had? And if natural selection cannot start a single organ of a single type, what is the use of discussing its supposed ability to improve them after the machinery is all built?

II

Such was the general condition of theoretical biology about the beginning of the present century. In the meantime those who were dealing with the empyrical or experimental side of these problems were seeking for the causes of and the rules for variation. All living things vary from one generation to another; the question was, Why do they vary? and do these variations really represent new characters comparable to new species in the making? or are they, so to speak, but an elastic reaction of the internal vital elasticity of the organism, all the while latent and only seeking a favorable expression, to return again under other conditions to the former type?

The effort to reduce these variations to law and system was pursued by thousands of investigators, with varying but at all times perplexing and disappointing results. But in the year 1900 the scientific world awoke to the surprising fact that a patient obscure investigator had already solved most of the puzzles of variation and heredity some thirty-five years before. Gregor Mendel, born a peasant boy, trained as a monk, and afterwards appointed Abbot of Bran, had in the year 1865 published the results of his experiments in breeding, which had been ignored or forgotten until rediscovered in 1900 by de Vries and two others simultaneously. From this point Mendelism, as it is now called, has steadily gained ground, until at the present time it can be said to be the dominating conception among biologists the world over regarding the problems of heredity.

Mendel worked chiefly with peas, crossing different varieties. In his methods of investigation he differed from all previous investigators in concentrating his attention upon a single pair of alternative or contrasted characters at a time, and observing how these alternative characters are transmitted.

Thus when he crossed a tall with a dwarf, giving attention to this pair of contrasted characters alone, he found that all the first hybrid generation were talls, with no dwarfs and no intermediates. Accordingly he called the tall character _dominant_, and the dwarf character _recessive_, and a pair of contrasted characters which act in this way are now called factors or sometimes called unit characters. But on allowing these hybrids to cross-fertilize one another in the usual way, Mendel found that in the second generation of hybrids there were always three talls to one dwarf out of every four. Further experiments proved that these dwarfs of the second hybrid generation _always bred true_, that is, one out of four; and that one out of the remaining talls always bred true, making another quarter of the total; while the remaining fifty per cent. proved to be mixed tails, always acting as did the original hybrids, splitting up in the next generation in the same arithmetical proportion as before.

Accordingly, if we confine our study to the two contrasted characters, tallness and dwarfness, we see that just three kinds of peas exist, namely, dwarfs which breed true, talls which breed true, and talls which always give the same definite proportion of talls and dwarfs among their descendants. Innumerable experiments which have since been made with other pairs of characters have demonstrated that this same mathematical proportion holds good throughout the whole world of plants and animals;[25] and hence this astonishing result is now called Mendel's Law, and is regarded as the most important discovery in biology in several generations.

[Footnote 25: When dealing with only a few individual cases, we do not always find them to come out in such exact proportion; but when the number of examples is large, the proportion is so close to these figures that the exceptions can be entirely neglected as probably due to error of some kind.]

There are two distinct kinds of Andalusian fowls, one pure bred black, the other pure bred white with slight dashes of black here and there. When these are mated, no matter which color is the father or the mother, the next or hybrid generation are always a queer mixture of black and white called by fanciers blue. When these blues are interbred, one-quarter of their offspring will be white, which will prove to breed true ever afterwards, one-quarter will be black that will breed true, and fifty per cent. will be blue which will break up in the next generation in the very same way as before. In this case neither white nor black character is dominant, and accordingly we have a blending of both in the first hybrid generation.

In guinea pigs, black color has been found to be dominant over white, rough coat over smooth coat, and short hair over long hair. These remarkable results following from an experimental trial of Mendelism have stimulated hosts of investigators in all parts of the world, until now many varieties of plants and animals have been studied for many successive generations, already, building up a considerable literature dealing with the subject.

Perhaps the most extensive and exact series of experiments along this line have been carried on by Thomas Hunt Morgan and his assistants, of Columbia University. For over five years they have been breeding the wild fruit fly (_Drosophila ampelophila_), during which time they have originated and observed over a hundred and twenty-five new types that breed true according to Mendel's laws. Every part of the body has been affected by one or another of these mutations. The wings have been shortened, or changed in shape, or made to disappear entirely. The eyes have been changed in color or entirely eliminated. And each of these wonderful variations was brought about not gradually, but at a single step.

Professor Morgan grows justifiably sarcastic in contrasting these demonstrated laboratory facts with the armchair theories that have so long and so harmfully dominated biological studies. A quotation from him will not be out of place at this point.

"I may recall in this connection that wingless flies also arose in our cultures by a single mutation. We used to be told that wingless insects occurred on desert islands because those insects that had the best developed wings had been blown out to sea. Whether this is true or not, I will not pretend to say; but at any rate wingless insects may also arise, not through a slow process of elimination, but at a single step.... Formerly we were taught that eyeless animals arose in caves. This case shows that they may also arise suddenly in glass milk bottles, by a change in a single factor."[26]

[Footnote 26: "A Critique of the Theory of Evolution," p. 67.]

We need not be particularly concerned here with the theoretical explanations of these facts offered in terms of the microscopic or even the infra-microscopic components of the germ cells. Morgan seems to make out a strong case for the theory that the chromosomes found in the nucleus are the real ultimate units that carry the hereditary factors. But he is quite decided in the opinion that these hereditary factors are fixed, and are not changed from generation to generation either by environment or by selection.[27] The important thing for us in this connection is to get a clear idea of the results following from an application of Mendel's laws to the old, old problem of the origin of species, incidentally noticing how the theory associated with Darwin's name now looks in the light of these new facts.

[Footnote 27: In human beings it has been found that the effects of alcoholism and of syphilis are indeed transmitted according to Mendelian law, being the two solitary examples of diseased conditions that are thus transmitted. But they are so plainly pathologic phenomena that there is little temptation for the advocates of Lamarckianism to use them as proofs of their theory.]

We have hitherto been considering the results worked out by Mendel with but one pair of contrasted characters or factors. But Mendel studied the relation of other characters of the pea, and found among other results that smooth seeds are dominant to wrinkled seeds, colored seeds dominant to white, yellow color dominant to green, etc. But when a combination of two factors in each parent are put into contrast by cross breeding, two wholly original forms (as they seemed) were sometimes produced, and it looked as if these new kinds were really analogous to new species.

For example, he crossed tall yellow peas with dwarf green peas, with the result that the first hybrid generation turned out to be all tall yellows. However, in the second hybrid generation they split up according to the law as already stated, modified by the additional complication brought into the problem by the additional pair of factors. For out of every sixteen plants there were nine tall yellows, three _dwarf yellows_, three _tall greens_, and one dwarf green. It is evident that these tall greens and dwarf yellows are really new forms; and further experiments proved that they can be separated out or segregated and grown as pure forms which thereafter breed true. Thus we have a very important result for the breeder, for it enables him to work to a definite aim and combine certain desirable characters into a single form.

The term _mutation_, as already intimated, has been given to this process of producing new varieties in this way. The kinds so produced are termed _mutants_, and at first they were hailed by enthusiastic scientists as "elementary species." De Vries in particular gave much publicity to this idea; for he thought he had really produced a new kind comparable in every respect to a true species as produced by nature among wild plants. But the enthusiasm with which this applied result of Mendel's Law was at first hailed by biologists has gradually subsided; for it has been found that though these new forms will breed true under certain conditions, they are nevertheless _cross-fertile with the original forms_, and thus the circle can be completed back again by a return to the parent form, from which the new "species" can again be produced at will with the same mathematical exactness as before.

III

Where then are we?

Clearly we have not really produced any new species in any correct sense of the word. If we have produced new forms that breed true and that are seemingly just as deserving of the rank of distinct species as many now listed in scientific books, it only shows that our lists are sadly at fault, and that they are not all species that are called species. These experiments merely indicate that _the parent form possesses more potential characters than it can give expression to in a single individual form_, some of them being necessarily latent or hidden, and that when these latent ones show themselves they must do so at the expense of others which become latent or hidden in their turn. This _vital elasticity_, as it may be termed, or the vital rebound under definite conditions, is indeed a prime characteristic of the species just as it is of the individual; but like that of the individual the vital elasticity of the species is strictly bounded by comparatively narrow limits beyond which we have never seen a single type pass under either natural or artificial conditions. Mutations can be made according to Mendel's Law; but when we have made them once we can always be sure of producing the _very same mutants again in the very same way_, as surely as we produce a definite chemical compound; and when we have made it _we can always resolve it at will back into its original form_, just as we can a chemical compound. And so, where is the evolution? or how do these facts throw any light on the problem of the origin of species, any more than chemical compounds throw light on the origin of the elements? Obviously in biology as in chemistry we are only working in a circle, merely marking time.

And the bearing of these facts on the other problem of the transmission of acquired characters is quite obvious. Mendelism provides no place for any such transmission. Mendel's Law is sometimes called the law of _alternative inheritance_, thus embodying in its name the thought that offspring may show the characters possessed by one parent or by the other, but that it cannot develop any characters whatever which were not manifest or latent in the ancestry. Changes in the environment during the embryonic stage, it is true, seem sometimes to be registered in the growing form; but it has never yet been proved that these induced changes can ever amount to a unit character or genetic factor that will maintain itself and segregate as a distinct factor after hybridization. Ancestry alone furnishes the material for the factor, and no amount of induced change can get itself registered in the organism so as to come into this charmed circle of ancestral characters which alone seem to be passed on to posterity.

A quotation from Bateson ought to set this point at rest:

"The essence of the Mendelian principle is very easily expressed. It is, first, that in great measure the properties of organisms are due to the presence of distinct, detachable elements [factors], separately transmitted in heredity; and secondly, that _the parent cannot pass on to offspring an element, and consequently the corresponding property, which it does not itself possess_."[28]

[Footnote 28: Scientific American Sup., January 3, 1914.]

Heredity we now see is a method of analysis, and the facts brought to light by Mendelism help us very much toward an understanding of living matter. Especially does it help us to understand the complexity underlying the facts of heredity, which until now have seemed so strange and capricious. As Professor Punnett of Cambridge remarks:

"Constitutional differences of a radical nature may be concealed beneath an apparent identity of external form. Purple sweet peas from the same pod, indistinguishable in appearance and of identical ancestry, may yet be fundamentally different in their constitution. From one may come purples, reds, and whites; from another only purples and reds; from another purples and whites alone; whilst a fourth will breed true to purple. Any method of investigation which fails to take account of the radical differences of constitution which may underlie external similarity, must necessarily be doomed to failure. Conversely, we realize to-day that individuals identical in constitution may yet have an entirely different ancestral history. From the cross between two fowls with rose and pea combs, each of irreproachable pedigree for generations, come single combs in the second generation, and these singles are precisely similar in their behavior to singles bred from strains of unblemished ancestry. In the ancestry of the one is to be found no single over a long series of years; in the ancestry of the other nothing but singles occurred. The creature of given constitution may often be built up in many ways, but once formed it will behave like others of the same constitution."[29]

[Footnote 29: Encyclopedia Britannica, Vol. XVIII, p. 119.]

IV

Vanished at last are the old theories of gradual changes in species perpetuated and accumulated by natural selection until at last wholly new forms have in this way been produced. True variations are now seen to be confined within well-marked and rather narrow limits, within which ordinary variations may occur, perhaps induced by environment. These fluctuating variations grade off into one another on all sides, and their differences can be plotted on a frequency curve; but the very important thing for us to remember is that these fluctuating variations _cannot be transmitted._ Beyond these fluctuating variations come the unit characters or factors, which are distinct from each other, or "discontinuous," to use the technical term, and which therefore cannot be plotted on a frequency curve. These factors are not modified in the least by the environment, and their peculiarities are faithfully transmitted in heredity with all the precision of chemical law. But even these factors are all within the bounds of the species. There is not a shred of scientific evidence that either natural or artificial devices have originated a single genetic factor that was not all the time potentially latent in the ancestry, capable of being produced at will by the proper combination.

It is a universal law of living things that all forms left to themselves tend to degenerate. The necessity for continuous artificial selection in the sugar beet, in Sea Island cotton, in corn, in Jersey and Holstein cattle, in trotting horses, proves this universal tendency to degenerate.[30] Natural selection in a somewhat similar way tends to postpone this degeneracy by killing off the "unfit," but selection either artificial or natural cannot originate anything new, and its results are here displayed merely among the small fluctuating variations mentioned above. Even among the real genetic factors it may show itself by allowing some to survive alone; but as no combination of diverse factors can originate anything really new, its field for operation among these factors is extremely limited. Among species also it is operative, killing off some and allowing others to survive. But neither among fluctuations, among factors, nor yet among species can selection originate anything new.

[Footnote 30: The following represents the consensus of scientific opinion regarding the lessons to be drawn from the phenomena of our improved races of domesticated plants and animals:

"One need not be a pessimist to assert the actual evidence thus far obtained indicates that the supposed progress made in the improvement of domesticated animals and plants is nothing more than the sorting out of pure lines, and thus represents no advancement."--Prof. L.B. Walton, _Science_, April 3, 1914.]

Nor is there any other method known to modern science by means of which new factors can be originated which were not potentially latent in the ancestry. The much heralded new "species" of de Vries and others are now known to be merely new factors cropping out;[31] for though they remain constant and breed true, they obey Mendel's Law when crossed with their parental forms, and hence are merely the result of some new combination of factors which can be reproduced at will by using the same method of combination and segregation. The real scientific test for any form supposed to be a new "species" would be twofold: (1) to show that some new character had been added which no ancestor ever possessed; and (2) to show that this new character will breed true under all circumstances of hybridization and not merely segregate as a unit character or mere analytic variety after hybridization. It is almost superfluous to say that no "new species" originating in modern times has ever justified itself under these tests.

[Footnote 31: Some of our leading biologists are now disposed to grow somewhat humorous when speaking of this mutation theory of de Vries, as may be illustrated by the following:

"The mutation theory of de Vries appears accordingly to lag useless on the biological stage, and may apparently be now relegated to the limbo of discarded hypotheses.... The present refutation has been undertaken in the interest of biological progress in this country. It is now high time, so far as the so-called mutation hypothesis, based on the conduct of the evening primrose in cultures, is concerned, that the younger generation of biologists should take heed lest the primrose path of dalliance lead them imperceptibly into the primrose path to the everlasting bonfire."--Prof. Edw. C. Jeffrey (Harvard), in _Science_, April 3, 1914.]

In conclusion it may be remarked that biologists do not claim to have solved all the problems connected with heredity and variation. But the general results taught us by Mendelism are now established beyond controversy. Led by the German biologists, the leading scientists of the world had already acknowledged that "pure" Darwinism or natural selection cannot explain the origin of new organs or new forms. And now Mendelism destroys the other supposed foundation for biological evolution, by showing that small variations cannot be accumulated into large differences equal in value to a unit character or a new species. Thus the whole foundation of biological evolution has been completely undermined by these new discoveries; and were it not for the wide-spread credence the evolutionary theory has already received, and the intellectual momentum it has acquired tending to carry it on by its inertia into the future, it could be only a very short time now before the elaborate treatises attempting to orientate with it all the facts of religion and history would have to be consigned to the shelves labeled, "Of Historic Interest." For as Bateson remarked in his recent address as President before the British Association at Melbourne, Australia, the new knowledge of heredity shows that whatever evolution there is occurs by loss of factors and not by gain, and that in this way the progress of science is "destroying much that till lately passed for gospel."[32]

[Footnote 32: In commenting on these views of Bateson, Prof. S.C. Holmes, of the University of California, well speaks of them as "an illustration of _the bankruptcy of present evolutionary theory."--Science_, September 3, 1915.]

V

Let us sum up the situation. We began this chapter with the question, Have new kinds of plants and animals originated in modern times comparable in all essential respects with the idea of true species?

The answer of modern science is reluctantly obtained, but it is a negative. De Vries and others have indeed originated new kinds that were loudly hailed as new species, and are doubtless as deserving of specific rank as many already listed for years in the treatises of specialists. Indeed there is every reason to believe that almost countless numbers of our taxonomic species have originated from common ancestral originals. But as these so-called species are now known to be freely or moderately cross fertile with other related species, their hybrids following the ordinary laws of Mendelian inheritance, we see that they are not true species but mere analytic varieties.

In short, we now know that our taxonomic classifications have been marked off on altogether too narrow lines. This has tended greatly to confuse the question at issue. But from our enlarged views of the laws and nature of heredity and variation, as well as from the original intent of the term species as defined by the great scientist who originated it, the verdict of an impartial investigator must be that we have never seen a new species originate by any natural or artificial method since the dawn of scientific observation.

Here again we find the record of Creation confirmed; for the failure of the thousands of modern investigators to originate genuine new species proves that in this respect also Creation is not now going on. And all the analogies from the origin of matter, of energy, of life, and from the laws of the reproduction of cells, indicate that we have at last found rock bottom truth regarding the vexed question of the origin of species. So far as science can observe and record, each living thing on earth, in air, in water, reproduces "after its kind."

VII

GEOLOGY AND ITS LESSONS

I

In all the previous chapters I have not been giving any very new facts or any discoveries of my own. True, my conclusions from the facts may seem novel; but in general I have been giving merely facts which are almost universally acknowledged by educated men. The conservation laws of matter and of energy, the impassable gulf between the living and the not-living, the laws governing cell multiplication, are matters of common knowledge and will be found in the appropriate college text-books throughout the civilized world. Even the facts which I have presented regarding variation and heredity are admitted in one way or another by practically all biologists. But in following our general subject into the field of geology, I shall be obliged to present some comprehensive truths and general conclusions which are not so widely acknowledged, because only recently brought to light. However, as these facts and conclusions may seem very new and strange to many, I shall endeavor to build up my argument wholly on the recorded observations of the very highest authorities rather than on my own unsupported testimony; though for the sake of brevity I shall be obliged to refer the reader to my "Fundamentals of Geology" (1913) for some of the details.

One of the great outstanding ideas of geology as usually taught is that life has been on the globe for many millions of years, that in fact there has been a graded succession of different types of life in a well defined invariable order, from the lower and more generalized to the higher and more specialized. Quite obviously this succession of life was antagonistic to the former views of a literal Creation; and only on this supposed fact as an outline has the modern theory of biological evolution been built up. For if geology cannot furnish the most unquestionable proof that life has occurred in a very definite and invariable order, what is the use of talking about the development of one form of life into another by a gradual process of evolution?

One of the highest scientific authorities in America, Prof. Thomas Hunt Morgan, of Columbia University, has recently said, "The direct evidence furnished by fossil remains is by all odds the strongest evidence that we have in favor of organic evolution."[33] Accordingly we purpose to examine carefully what this by all odds "strongest evidence" is like.

[Footnote 33: "A Critique of the Theory of Evolution," p. 24.]

II

As with some of the other facts with which we have had to deal in previous chapters, a correct understanding of the questions involved can best be obtained by examining the history of the development of the science.

The first man with whom we need to concern ourselves is A.G. Werner, a teacher of mineralogy in the University of Freiberg, Germany. For three hundred years his ancestors had been connected with mining work, and he, though possessing little general education, knew about all that was then known regarding mineralogy and petrology. He wrote no books; but by his enthusiastic teaching he gathered as students and sent out as evangelists hundreds of devoted young scientists who rapidly spread his theories through all the countries of Europe.

"Unfortunately," says Zittel, "Werner's field observations were limited to a small district, the Erz Mountains and the neighboring parts of Saxony and Bohemia. And his chronological scheme of formations was founded on the mode of occurrence of the rocks within these narrow confines."[34]

[Footnote 34: "History of Geology," p. 59.]

Werner had found the granites, limestones, sandstones, schists, etc., occurring in a certain relative order in his native country; and he drew the very remarkable conclusion that this was the normal order in which these various rocks would invariably be found in all parts of the world, on the theory that this was the order in which these different rocks had been formed in the beginning, great layers of these different rocks having originally been spread completely around the globe one outside another like the coats of an onion. With this as a major premise, it is not surprising that he and his enthusiastic disciples "were as certain of the origin and sequence of the rocks as if they had been present at the formation of the earth's crust."[35]

[Footnote 35: A. Geikie, "Founders of Geology," p. 112.]

The amusement with which this onion-coat theory is now regarded is hardly appropriate in view of its universal vogue among geologists about the beginning of the nineteenth century, and in view of the further fact that a very similar and only slightly modified substitute theory has been universally taught for three-quarters of a century and still prevails. The modern form of the theory substitutes onion-coats of fossiliferous rocks for onion-coats of mineral and lithological characters; and a brief consideration of this theory is now in order.

About the time that various geologists here and there were finding rocks in positions that could not be explained in terms of Werner's theory, William Smith (1769-1839) in England and the great Baron Cuvier (1769-1832) in France found characteristic fossils occurring in various strata; and under their teachings it was not long before the fossils were considered the best guide in determining the relative sequence of the rocks. The familiar idea of world-enveloping strata as representing successive ages was not discarded; but instead of Werner's successive ages of limestone making, sandstone making, etc., these new investigators taught that there were successive ages of invertebrates, fishes, reptiles, and mammals, these creatures having registered their existence in rocky strata which thus by hypothesis completely encircled the globe one outside another.

It is true that early in the nineteenth century Sir Charles Lyell and others tried to disclaim this absurd and unscientific inheritance from Werner's onion-coats; but modern geology has never yet got rid of its essential and its chief characteristic idea, for all our text-books still speak of various successive ages when only certain types of life prevailed all over the globe. Hence it is that Herbert Spencer caustically remarks: "Though the onion-coat hypothesis is dead, its spirit is traceable, under a transcendental form, even in the conclusions of its antagonists."[36] Hence it is that Whewell, in his "History of the Inductive Sciences," refuses to acknowledge that in geology any real advance has yet been made toward a stable science like those of astronomy, physics, and chemistry. "We hardly know," he says, "whether the progress is begun. The history of physical astronomy almost commences with Newton, and few persons will venture to assert that the Newton of geology has yet appeared."[37] Hence it is that T.H. Huxley declares, "In the present condition of our knowledge and of _our methods_, one verdict,--'_not proven and not provable'--must be recorded against all grand hypotheses of the paleontologist respecting the general succession of life on the globe."[38] And hence it is that Sir Henry H. Howorth, a member of the British House of Commons and the author of three exhaustive works on the Glacial theory, declares, "It is a singular and notable fact, that while most other branches of science have emancipated themselves from the trammels of metaphysical reasoning, the science of geology still remains imprisoned in a priori theories."[39]

[Footnote 36: "Illustr. of Univ. Prog.," p. 343.]

[Footnote 37: Vol. II, p.580.]

[Footnote 38: "Discourses," pp. 279-288.]

[Footnote 39: "The Glacial Nightmare," Preface, vii.]

And thus the matter remains even to-day, in this second decade of the twentieth century. _Geology has never yet been regenerated_, as have all the other sciences, by being delivered from the caprice of subjective speculations and a priori theories and being placed on the secure basis of objective and demonstrable fact, in accordance with the principles of that inductive method of investigation which was instituted by Bacon and which has become so far universal in the other sciences that it is everywhere known as the scientific method. In accordance with this method, theories in all the other sciences are always kept well subordinated to facts; and whenever unequivocal facts are found manifestly contradicting a theory no matter how venerable, the theory must go to make way for the facts. In other words, the theoretical parts of the various other sciences are always kept revised from time to time, to keep them in line with the new discoveries that have been made. There has been no lack of astonishing discoveries of new facts in geology during the past half century or so, while all the other sciences have been making such astonishing progress. _But for over seventy five years geology has not made a single advance movement in its theoretical aspects_; indeed, in all its important general principles it has scarcely changed in a hundred years. I shall leave it to the reader to judge whether this is a case of almost miraculous perfection from the beginning, or of arrested development.

III

Of the three general postulates or a priori assumptions of this curiously out-of-date medi 鑒 al science, namely, (1) Uniformity, (2) the Cooling globe theory, and (3) the theory of the Successive Ages, the first two have already been examined and found wanting by other investigators, and have been allowed to lapse into a sort of honored disuse, though their memory is still reverently cherished in all the text-books of the science. The "Challenger" Expedition dissipated most of the myths that had long been taught regarding the deep waters of the ocean; and Professor Suess has disposed of the closely related myth about the coasts of the continents being constantly on the seesaw up and down. These two discoveries, with others that might be mentioned, dispose of Lyell's theory of uniformity. Lord Kelvin and the other physicists dissipated the idea of a molten interior of the earth. Hence, because these other false hypotheses have already in a measure been disposed of, as well as for the sake of brevity, I shall here discuss only the third of the prime postulates of the current system of geology, namely the theory of Successive Ages. And when we have adjusted this aspect of the science of geology to the facts of the rocks as made known to us by modern discoveries, we shall find little in this science out of harmony with the older view of a literal Creation as taught in the Bible and as already confirmed by the other branches of science which we have been examining.

There are five leading arguments against the reality of these successive ages. Four of them must be dismissed here by a brief summary of the facts as we know them to-day, referring the reader to the author's larger work, where detailed evidence is given for each. The fifth series of facts I shall give here in more detail, though of course even this must be but an outline of what is given elsewhere.

1. In the earlier days of the theory of successive ages it was taught that only certain kinds of fossils were to be found at the bottom of the series, or next to the Primitive or Archan. This feature of the theory was demanded by the supposed universal spread of one type of life all around the globe in the earliest age. But it is now known that the so-called "oldest" fossiliferous rocks occur only in detached patches over the globe, while other or "younger" kinds are just as likely to be found on the Primitive or next to the Archan. Not only may any kind of fossiliferous rocks occur next to the Arch鎵 n, but even the "youngest" may be so metamorphosed and crystalline as to resemble exactly in this respect the so-called "oldest" rocks. On the other hand some of the very "oldest" rocks may, like the Cambrian strata around the Baltic and in some parts of the United States, consist of "muds scarcely indurated and sands still incoherent."[40]

[Footnote 40: J.A. Howe; Encyclopedia Britannica, Vol. II, p. 86. Cambridge Edition.]

All this means that many facts regarding the position of the strata as well as regarding their consolidation contradict the theory of successive ages.

2. Many of the rivers of the world completely ignore the alleged varying ages of the rocks in the different parts of their course, and treat them all as if of the same age or as if they began sawing at them all at the same time. This is true of the Rhine, the Meuse, and the Danube in Europe, the Sutlej of India, and the upper part of the Colorado in America, not to mention others. The old strand lines around all the continents act in the very same way, ignoring the varying ages of the rocks they happen to meet; as is also true of nearly all the great faults or fissures which are of more than local extent. The ore veins of the various minerals are about as likely to be found in Tertiary or Mesozoic as in the Paleozoic. A very similar lesson is to be learned from the fossils found lying exposed on the deep ocean bottom; for they are about as likely to be Paleozoic or Mesozoic as Tertiary.

From these facts we conclude that practically all the great natural chronometers of the earth seem to treat the fossiliferous rocks as if they are _all of about the same age_, completely disregarding the distinctions in age founded on the fossils.

3. According to the present chronological arrangement of the rocks, very many genera, often whole tribes of animals, are found as fossils only in the oldest rocks, and _have skipped all the others_, though found in comparative abundance in our modern world. Very many others have skipped from the Mesozoic down, while still others skip large parts of the series of successive ages.

These absurdities would all be avoided by acknowledging that the current distinctions as to the ages of the fossils are purely artificial, and that one fossil is intrinsically just as old or as young as another.

4. It is now known that any kind of "young" beds whatsoever, Mesozoic, Tertiary, or even Pleistocene, may be found in such perfect conformability on some of the very oldest beds over wide stretches of country that "the vast interval of time intervening is unrepresented either by deposition or erosion"; while in some instances these age-separated formations so closely resemble

one another in structure and in mineralogical make-up that, "were it not for fossil evidence, one would naturally suppose that a single formation was being dealt with" (McConnell); and these conditions are "not merely local, but persistent over wide areas" (A. Geikie), so that the "numerous examples" (Suess) of these conditions "may well be cause for astonishment" (Suess).

A still more astonishing thing from the standpoint of the current theories is that these conformable relations of incongruous strata are often _repeated over and over again in the same vertical section_, the same kind of bed reappearing alternately with others of an entirely different "age," that is, appearing "as if _regularly interbedded"_ (A. Geikie) with them, in a manifestly undisturbed series of strata.

Here again we have a very formidable series of facts whose gravamen is directed wholly against the artificial distinctions in age between the different groups of fossils; and their argument is an eloquent plea that the fossils are neither older nor younger but all of a similar age.

5. Our last fact demands a somewhat more extended consideration; but it may be stated in advance briefly as follows:

In very numerous cases and over hundreds and even thousands of square miles, the conformable conditions specified in the previous fact are exactly reproduced _upside down_; that is, very "old" rocks occur with just as much appearance of natural conformability on top of very "young" rocks, the area in some instances covering many hundreds of square miles, and in one particular instance in Montana and Alberta covering about five or six thousand square miles of area.

The first notable example of this phenomenon was discovered at Glarus, Switzerland, a good many years ago; since which time this locality has become a classic in geological literature, and has called out many ponderous monographs in German and French by such men as Heim, Schardt, Lugeon, Rothpletz, and Bertrand. This example, which was first (1870) called the

Glarner Double Fold by Escher and Heim, is now universally called a nearly flat-lying "thrust fault," in accordance with the explanations since adopted of similar phenomena elsewhere. Without obtruding unnecessary technicalities upon my non-professional readers, I may quote the words of Albert Heim as to the conditions as now recognized in these parts:

"These flat-lying faults, of which those at Glarus were the first to be discovered, are a universal phenomenon in the Northern and Central Alps."[41]

[Footnote 41: "Der Bau der Schweizeralpen," p. 17.]

The favorite method of explaining these conditions has slightly changed within recent years, as already remarked. For whereas the classic example at Glarus was at first spoken of as a double fold-in from both sides toward the Sernf Valley, this is now universally spoken of as a "thrust fault," with the rocks all pushed one way. Incidentally it may be noted that this very fact that what was long regarded as two completely overturned folds is now spoken of as one flat-lying thrust fault, is prima facie evidence that there is here no physical proof of any real overturning of the strata, such as we do find on a very small scale in true folded rocks. The latter can usually be measured in yards, feet, or inches; while in this example at Glarus the area involved would be measured in many miles, and in some very similar examples to be presently mentioned from America the measurement could best be made in degrees of latitude and longitude or in arcs of the earth's circumference. In these larger examples it is manifestly impossible that there should be any physical evidence sufficient to indicate a huge earth movement of this character, especially when, as is usually the case, both the upper and the lower strata are quite uninjured in appearance. No; the fossils are here in the wrong order, that is all. And so, to save the long established doctrines of a very definite order of successive life-forms, this theory of a "thrust fault" is offered as the best available explanation. As Dr. Albert Heim himself once expressed it very naively in a letter to the present writer, that the strata over these large areas are in a position manifestly at direct disagreement with the received order of the fossils, "is a fact which can be clearly seen,--only we know not yet how to explain it in a mechanical way."

An example in the Highlands of Scotland was about the next to be discovered. Here, as Dana says, "a mass of the oldest crystalline rocks, many miles in length from north to south, was thrust at least ten miles westward over younger rocks, part of the latter fossiliferous;" and he further declares, "the thrust planes _look like planes of bedding, and were long so considered._"[42]

Sir Archibald Geikie and others had at first described these beds as naturally conformable; and when at length they were convinced that the fossils would not permit this explanation, Geikie gives us some very picturesque details as to how natural they look.

The thrust planes, he says, are with much difficulty distinguished "from ordinary stratification planes, like which they have been plicated, faulted, and denuded. Here and there, as a result of denudation, a portion of one of them appears capping a hilltop. One almost refuses to believe that the little outlier on the summit does not lie normally on the rocks below it, but on a nearly horizontal fault by which it has been moved into its place."

Of a similar example in Ross Shire he declares:

"Had these sections been planned for the purpose of deception, they could not have been more skilfully devised, ... and no one coming first to the ground would suspect that what appears to be a normal stratigraphical sequence is not really so."[43]

[Footnote 42: "Manual," pp. 111, 534.]

[Footnote 43: _Nature_, November 13, 1884, pp. 29-35.]

Here again we have unequivocal testimony from the most competent of observers that there is no physical evidence whatever to lead any one to say that a ponderous scale of the earth's crust was really pushed up on top of other portions, as this makeshift theory of "thrust faults" involves. The _fossils are here in the wrong order_, just as in the case at Glarus; that is all. The facts seem to be a flat contradiction to the theory of definite successive ages, and to save the theory this explanation of a "thrust fault" is invented, though there is absolutely no physical evidence of any disturbance of the strata.

Our next stopping place is in the Southern Appalachian Mountains of eastern Tennessee and northern Georgia. Here we have the Carboniferous strata dipping gently to the southeast, like an ordinary low monocline, under Cambrian or Lower Silurian, one of these so-called faults having a reported length of 375 miles,[44] while in another instance the upper strata are said to have been pushed about eleven miles in the direction of the "thrust."[45] These conditions, we are told, "have provoked the wonder of the most experienced geologists,"[46] because of the perfectly natural appearance of the surfaces of the strata affected; or as this same writer puts it, "The mechanical effort is great beyond comprehension, but the effect upon the rocks is inappreciable," and "the fault dip is often parallel to the bedding of the one or the other series of strata."[47] Which means, in other words, that these "thrust planes" look just like ordinary planes of bedding between conformable strata.

[Footnote 44: Bailey Willis, Geol. Survey, Report, Vol. 13, p. 228.]

[Footnote 45: C.W. Hayes, _Bull. Geol. Soc_., Vol. 2, pp. 141-154.]

[Footnote 46: Willis, _op. cit_., p. 228.]

[Footnote 47: Willis, _op. cit_., p. 227.]

The Rocky Mountains furnish examples of many kinds of natural phenomena on the very largest scale, and those of the sort here under consideration are no exception to this rule. For here we have an immense area east of the main divide, extending from the middle of Montana up to the Yellowhead Pass in Alberta, or over 350 miles long, where the tops of the mountains consist of jointed limestones or argillites of Algonkian or pre-Cambrian "age," resting on soft Cretaceous shales. Often the greater part of the mass of a range will consist of these "older" and harder rocks, which by the erosion of the soft underlying shales are left standing in picturesque, rectangular, cathedral-like masses, easily recognizable as far off as they can be seen. And the almost entire absence of trees or other vegetation helps one to trace out the relationship of these formations over immense areas with little or no difficulty.

In the latitude of the Bow River, near the Canadian Pacific main line, there is a long narrow valley of these Cretaceous beds some sixty-five miles long, called the Cascade Trough, with of course pre-Cambrian mountains on each side. Somewhat further south there are two of these Cretaceous valleys parallel to one another, and in some places _three_; while just south of the fiftieth parallel of latitude, at Gould's Dome, there are actually five parallel ranges of these Paleozoic mountains, _with four Cretaceous valleys in between_, one of these valleys, the Crow's Nest Trough, being ninety-five miles long.

But we ought to take a nearer view of these wonderful conditions. A convenient point of approach will be just east of Banff, Alberta, near Kananaskis Station, where the Fairholme Mountain has been described by R.G. McConnell of the Canadian Survey. The latter remarks with amazement on the perfectly natural appearance of these Algonkian limestones resting in seeming conformability on Cretaceous shales, and says that the line of separation between them, called in the theory the "thrust plane," resembles in all respects an ordinary stratification plane. I quote his language:

"The angle of inclination of its plane to the horizon is _very low_, and in consequence of this its outcrop follows a very sinuous line along the base of the mountains, and acts exactly like the line of contact of two nearly horizontal formations.

"The best places for examining this fault are at the gaps of the Bow and of the south fork of Ghost River.... The fault plane here is nearly horizontal, and the two formations, viewed from the valley, _appear to succeed one another conformably._"[48]

[Footnote 48: Annual Report, 1886, Part D, pp. 33, 34.]

This author adds the further interesting detail that the underlying Cretaceous shales are "very soft," and "have suffered very little by the sliding of the limestone over them."[49]

About a hundred miles further south, but still in Alberta, we have the well-known Crow's Nest Mountain, a lone peak, which consists of these same Algonkian limestones resting on a Cretaceous valley "in a nearly horizontal attitude," as G.M. Dawson says, which "in its structure and general appearance much resembles Chief Mountain,"[50] another detached peak some fifty miles further south, just across the boundary line in Montana.

Chief Mountain has been well described by Bailey Willis,[51] who estimates that the Cretaceous beds underneath this mountain must be 3,500 feet thick; while the so-called "thrust plane is essentially _parallel to the bedding_" of the upper series.[52]

"This apparently is true not only of the segments of thrust surface beneath eastern Flattop, Yellow, and Chief Mountain, but also of the more deeply buried portions which appear to dip with the Algonkian strata into the syncline. While observation is not complete, it may be assumed on a basis of fact that thrust surfaces and bedding are nearly parallel over extensive areas."[53]

[Footnote 49: Report, 1886, Part D, p. 84.]

[Footnote 50: Report, 1885, Part B, p. 67.]

[Footnote 51: _Bull. Geol. Soc._, Vol. 13, pp. 305-352.]

[Footnote 52: Id., p. 336.]

[Footnote 53: Id., p. 336.]

Quite recently this region has been studied by Marius R. Campbell of the Washington Survey Staff (Bulletin 600), while the part in Alberta has been studied by Rollin T. Chamberlin of Chicago. Much of the vast area involved is not yet well explored; but over it all, so far as it has been fully examined, the same lithological and stratigraphical structures reappear with the persistence of a repeating decimal. And were it not for the exigencies of the theory of Successive Ages, this whole region of some five or six thousand square miles would be considered as only an ordinary example, on a rather large scale, of undisturbed horizontal stratification cut up by erosion into mountains of denudation, with of course occasional instances of minor local disturbances here and there, as would be expected over an area of this extent.

Richards and Mansfield in a recent paper describe the "Bannock Overthrust," some 270 miles long, in Utah, Idaho, and Wyoming. The Carnegie Research recently reported a similar phenomenon about 500 miles long in northern China.

But it would be tiresome to follow these conditions around the world. We have plenty of examples, and we have them described by the foremost of living geologists. What we need to do now is to adopt a true scientific attitude of mind, a mind freed from the hypnotizing influence of the current theories, in order correctly to interpret the facts as we already have them.

How much of the earth's crust would we have to find in this upside down order of the fossils, before we would be convinced that there must be something hopelessly wrong with this theory of Successive Ages which drives otherwise competent observers to throw away their common sense and cling desperately to a fantastic theory in the very teeth of such facts?

The science of geology as commonly taught is truly in a most astonishing condition, and doubtless presents the most peculiar mixture of fact and nonsense to be found in the whole range of our modern knowledge. In any minute study of a particular set of rocks in a definite locality, geology always follows facts and common sense; while in any general view of the world as a whole, or in any correlation of the rocks of one region with those of another region, it follows its absurd, unscientific theories. But wherever it agrees with facts and common sense, it contradicts these absurd theories; and wherever it agrees with these theories, it contradicts facts and common sense. That most educated people still believe its main thesis of a definite age for each particular kind of fossil is a sad but instructive example of the effects of mental inertia.

IV

The reader will find this matter discussed at length in the author's "Fundamentals of Geology"; but here it will be necessary only to draw some very obvious conclusions from the five facts which we have set in opposition to the theory of Successive Ages.

1. The first and absolutely incontrovertible conclusion is that this theory of successive ages must be a gross blunder, in its baleful effects on every branch of modern thought deplorable beyond computation. But it is now perfectly obvious that the geological distinctions as to age between the fossils are fantastic and unjustifiable. No one kind of true fossil can be proved to be older or younger than another intrinsically and necessarily, and the methods of reasoning by which this idea has been supported in the past are little else than a burlesque on modern scientific methods, and are a belated survival from the methods of the scholastics of the Middle Ages.

Not by any means that all rock deposits are of the same age. The lower ones in any particular locality are of course "older" than the upper ones, that is, they were deposited first. But from this it by no means follows that the fossils contained in these lower rocks came into being and lived and died before the fossils in the upper ones. The latter conclusion involves several additional assumptions which are wholly unscientific in spirit and incredible as matters of fact, one of which assumptions is the _biological form of the onion-coat theory_. But since thousands of modern living kinds of plants and animals are found in the fossil state, _man included_, and no one of them can be proved to have lived for a period of time alone and before others, we must by other methods, more scientific and accurate than the slipshod methods hitherto in vogue, attempt to decide as best we can how these various forms of life were buried, and how the past and the present are connected together. But the theory of definite successive ages, with the forms of life appearing on earth in a precise and invariable order, is dead for all coming time for every man who has had a chance to examine the evidence and has enough training in logic and scientific methods to know when a thing is really proved.

And how utterly absurd for the friends of the Bible to spend their time bandying arguments with the evolutionist over such minor details as the question of just what geological "age" should be assigned for the first appearance of man on the earth, when the evolutionist's major premise is itself directly antagonistic to the most fundamental facts regarding the first chapters of the Bible, and above all, when this major premise is really the weakest spot in the whole theory, the one sore spot that evolutionists never want to have touched at all.

I fancy I hear some one object, and ask what we are to do with the systematic arrangement of the fossils, the so-called "geological succession," that monument to the painstaking labors of thousands of scientists all over the world. This geological series is still on our hands; what are we to do with it?

It is scarcely necessary for me to say that this arrangement of the fossils is not at all affected by my criticism of the cause of the geological changes. _The geological series is merely an old-time taxonomic series, a classification of the forms of life that used_ _to live on the earth_, and is of course just as artificial as any similar arrangement of the modern forms of life would be.

We may illustrate the matter by comparing this series with a card index. The earlier students of geology arranged the outline of the order of the fossils by a rather general comparison with the series of modern life forms, which happened to agree fairly well with the order in which they had found the fossils occurring in England and France. But only a block out of the middle of the complete card index could be made up from the rocks of England and France; the rest has had to be made up from the rocks found elsewhere. Louis Agassiz did herculean work in rearranging and trimming this fossil card index so as to make it conform better, not only to the companion card index of the modern forms of life, but also to that of the embryonic series. From time to time even now readjustments are made in the details of all three indexes, the fossil, the modern, and the embryonic, the method of rearrangement being charmingly simple: just taking a card out of one place and putting it into another place where we may think it more properly belongs. And then if we can convince our fellow scientists over the world that our rearrangement is justified, our adjustment will stand,--until some one else arises to do a better job. When a new set of rocks is found in any part of the world it is simplicity itself for any one acquainted with the fossil index system to assign these new beds to their proper place, though of course the one doing this must be prepared to defend his assignment with pertinent and sufficient taxonomic reasons.

In view of these facts, we need not be concerned as to the fate of the geological classification of the fossils. It is a purely artificial system, just as is the modern classification; but both are useful, and so far as they represent true relationships they will both stand unaffected by any change we may make in our opinions as to how the fossils were buried. But in view of this purely artificial character of the geological series, what a strange sight is presented by the usual methods employed to "prove" the exact order in which evolution has taken place, such for instance as the use made of the graded series of fossil "horses," to illustrate some particular theory of just how organic development has occurred. One might just as well arrange the modern dogs from the little spaniel to the St. Bernard, for the geological series is just as artificial as would be this of the dogs.

2. Another conclusion from the facts enumerated above is that there has obviously been a great world catastrophe, and that this must be assigned as the cause of a large part,--just how large a part it is at present difficult to say,--of the changes recorded in the fossiliferous rocks. This sounds very much like a

modern confirmation of the ancient record of a universal Deluge; and I say confidently that no one who will candidly examine the evidence now available on this point can fail to be impressed with the force of the argument for a world catastrophe as the general conclusion to be drawn from the fossiliferous rocks all over the globe.

3. Finally, there is the further conclusion, the only conclusion now possible, if there is no definite order in which the fossils occur, namely, that life in all its varied forms _must have originated on the globe by causes not now operative_, and this Creation of all the types of life may just as reasonably have taken place all at once, as in some order prolonged over a long period.

As I have pointed out in my "Fundamentals," a strict scientific method may destroy the theory of Successive Ages, and it may show that there has been a great world catastrophe. But here the work of strict inductive science ends. It cannot show just how or when life or the various kinds of life did originate, it can only show how it did not. It destroys forever the fantastic scheme of a definite and precise order in which the various types of life occurred on the globe, and thus it leaves the way open to say that life must have originated by just such a literal Creation as is recorded in the first chapters of the Bible. But this is as far as it can be expected to go. It is strong evidence in favor of a direct and literal Creation; but it furnishes this evidence by indirection, that is, by demolishing the only alternative or rival of Creation that can command a moment's attention from a rational mind.

But if life is not now being created from the not-living, if new kinds of life are not now appearing by natural process, if above all we cannot prove in any way worthy of being called scientific that certain types of life lived before others, if in fine man himself is found fossil and no one fossil can be proved older than another or than that of man himself, why is not a literal Creation demonstrated as a scientific certainty for every mind capable of appreciating the force of logical reasoning?

VIII

CREATION AND THE CREATOR

I

We need not here attempt to discuss the existence or even the nature of God. The Infinite One in all His attributes is above and beyond discussion. But there are some things that we can very profitably gather together as the net results of modern scientific investigation regarding the origin of things; and to this task we must now address ourselves in a very brief way.

We shall not attempt to deal with the astronomical aspects of the question, or the origin of our world as a planet or the origin of the solar system. This would lead us too far afield. We shall make more progress in dealing with the questions nearest at hand, namely, the origin of the present order of things on our globe.

First we must summarize the facts as we now know them in the five departments of knowledge with which we have had to deal.

1. Both matter and energy seem now to be at a standstill, so far as creation is concerned; no means being known to science whereby the fixed quantity of both with which we have to deal in this world can be increased (or diminished) in the slightest degree.

2. The origin of life is veiled in a mist that science has not dispelled and does not hope to dispel. By none of the processes that we call natural can life now be produced from the not-living.

3. Unicellular forms can come only from preexisting cells of the same kind; and even the individual cells of a multicellular organism, when once differentiated, reproduce only other cells after their own kind.

4. Species of plants and animals have wonderful powers of variation; but

these variations seem to be regulated and predestined in accordance with definite laws, and in no instance known to science has this variation resulted in producing what could properly be called a distinct new kind of plant or animal.

5. Geology has been supposed to prove that there has been a long succession of distinct types of life on the globe in a very definite order extending through vast ages of time. This is now known to be a mistake. Most living forms of plants and animals are also found as fossils; but there is no possible way of telling that one kind of life lived and occupied the world before others, or that one kind of life is intrinsically older than any other or than the human race.

II

In view of such facts as these, what possible chance is there for a scheme of organic evolution?

Must we not say that every possible form of the development theory is hereby ruled out of court? There can be no thought of the gradual development of organic nature by every-day processes in a world where such facts prevail. Rather must we say, with the force of the accumulated momentum of all that has been won by modern science, that, instead of the animals and plants on our world having arisen by a long-drawn-out process of change and development of one kind into another, there must have been just such a literal Creation at the beginning as the Bible describes. As we stand with uncovered head and bowed form in the presence of this great truth, it would seem almost like sacrilege to attempt by rhetoric to adorn it. Its inevitableness, its majesty, its transcendent importance for our generation, would only be obscured by so doing.

The essential idea of the Evolution theory is uniformity. It seeks to show that the present orders of plant and animal life originated by causes or processes identical with those now said to be operating in our modern world. It denies that at any particular time in the past causes and processes were in operation to originate the present order of nature which were essentially different from the processes now operating in our world under what we call natural law. Evolution seeks to smooth out all distinction between Creation and the modern r 間 ime of "natural law."

On the other hand, the essential idea of the Christian doctrine of Creation is that, back at a period called "the beginning," forces and powers were brought into exercise and results were accomplished which have not since been exercised or accomplished. In other words, the origin of the world and the things upon it was essentially and radically different from the manner in which the present order of nature is now being sustained and perpetuated. The mere matter of time is in no way the essential idea in the problem. The question of how much time was occupied in the work of Creation is of no importance, neither is the question of how long ago it took place. The one essential idea is that the processes and methods of Creation are beyond us, for we have nothing with which to measure it; Creation and the reign of "natural law" are essentially incommensurable. The one thing that the doctrine of Creation insists upon is that the origin of our world and of the things upon it must have been brought about by some direct and unusual manifestation of the power of the Being whom we call the Creator; and that since this original Creation the things of nature have been perpetuated and sustained by processes and methods which (though still essentially inscrutable by us) we call the order of nature and the reign of natural law.

But in view of the series of facts enumerated in the previous pages, the doctrine of Creation is established by modern scientific discoveries almost like the conclusion of a mathematical problem.

III

How are modern intelligent men and women to avoid any longer this inevitable conclusion of a literal Creation as the method of origin for our world and the things upon it?

The facts enumerated in the previous pages are not new; it is only the present grouping and arrangement of them, and the conclusions drawn from them, that are new. Of all the leading facts enumerated above, only the last one, the one regarding geology, is any longer a subject of serious discussion by educated people. And the general facts as stated above regarding geology have been proved (by the present writer) with such a wealth of facts and arguments that they also must speedily be acknowledged by scientists, when the latter take the trouble to study these facts and arguments. And with geology once adjusted to a system of real inductive science, instead of being as hitherto under the hypnotic control of speculative fancies and subjective methods, there is no longer any room for speculations regarding the origin of our world by evolutionary processes. It becomes almost a mathematical Q.E.D. _that things were made in the beginning by methods and processes that are no longer operative_, so far as science can observe. This means a real Creation, in the Bible sense of the term, something distinct from the means by which nature is now being sustained and carried on. Any attempt to describe the why or the how of this Creation would be useless speculation; but _this much is science_, and science that is to-day all the more impressive and conclusive because it has been won by centuries of conflict with every conceivable opposing prejudice.

IV

In conclusion we may attempt to speak in a brief way of the present relationship between the Creator and the things which He has made, and if possible to dispel the sad confusion prevailing in many minds between God's continued immediate action in certain departments of nature and His action in other departments through the intermediate use of second causes.

On every hand we hear proclaimed a form of the doctrine of God's omnipresence (usually called the divine "immanence") which not only denies all distinction between the original Creation and the present perpetuation of the world, but a form which practically denies all second causes, and which cannot well be distinguished from pantheism, though it would be a spiritualistic or "idealistic" form of pantheism, or "monism," to use the favorite modern term. These extreme advocates of what they term the divine "immanence" go so far as to deny all second causes. And while they are fond of proclaiming this idea as an entirely new discovery, and proclaiming it with all the enthusiasm of proselytes to a new religion, they are also prone to state the (seemingly) opposed doctrine of second causes in such a way that it amounts to a mere caricature, a burlesque, picturing a sort of "absentee" God, who started the universe running and now merely stands by and watches it go. Thus pantheism and deism are often spoken of as the only alternatives for the choice of the modern man; for the real teachings of the Bible and of Christian philosophy are as completely ignored as if they had never been formulated or taught by intelligent people.

Let us first consider the scientific aspects of the doctrine of second causes, and the doctrine of God's immediate acting in various departments (or all departments) of nature.

1. We cannot deny that the will of man is a real cause, producing continual changes in the world about us. More than this, if there are not also second causes outside of the will of free intelligent personalities, the whole universe must be a gigantic deception; for it seems to be full of second causes. Long chains of what seem like second causes exist, made up of infinite numbers of links, as when the sun carries an amount of water up into the air, the latter dropping the water upon a mountain in the form of rain, gravity rolling it down the slope in vast force, sweeping away villages and towns, changing the fates of individuals and of nations. To quote two familiar examples from Stewart and Tait: "In a steam engine the amount of work produced depends upon the amount of heat carried from the boiler into the condenser; and this amount depends in its turn upon the amount of coal which is burned in the furnace of the engine. In like manner the velocity of the bullet which issues from the rifle depends upon the transformation of the energy of the powder; this in turn depends upon the explosion of the percussion cap; this again upon the fall of the trigger; and lastly this upon the finger of the man who fires the rifle."[54] Thus even the very strongest opponents of the idea of second causes never deny that the latter seem to surround us on every side, and that it would be

possible to trace a continuous line of apparent effects and causes back to the very beginning.

[Footnote 54: "The Unseen Universe," p. 184.]

This view of the matter, it is evident, readily leads to a deistic view of the universe,--or to that burlesque of the Christian view spoken of as the doctrine of an "absentee God," watching His universe run from the outside, slightly concerned with what it does.

2. On the other hand, a careful study of the correlation of forces shows us that the great First Cause is still very closely related to the operation of His universe. We may start, for instance, with the old argument from the evidences of design in nature, which, though often sneered at of late, cannot be cavalierly dismissed in this way; for, as Dugald Stewart has well said, "every combination of means to an end implies intelligence." But the direct or immediate action of the great Intelligence behind nature is manifest in the marvellous behavior of the cells; which, instead of behaving in a way to indicate that their life processes are due to properties inherent in the atoms and molecules composing them, show every appearance of being mere automata under the direct control of an intelligent, purpose-filled Mind,--a Mind external to themselves, it is true, and gloriously transcending them, but constantly, ceaselessly exercised by an immediate action which we may well call "immanent," in the original and proper sense of this term. Yet vital action is capable of exact correlation with the other forces of nature; and thus the modern law of the correlation of forces teaches us that the energy behind life must be the same as the energy pervading all nature, the various manifestations of which we know as light, heat, gravity, electricity, etc. Thus while the study of the behavior of life or the doctrine of "vitalism" might encourage us to think that in the cells and in the behavior of protoplasm we are witnessing the direct action of an intelligent Creator; yet we find that by the correlation of forces we must say the same about all the physical and chemical phenomena of nature. In other words, while the study of mere physical and chemical action might easily lead us to a strong belief in second causes, or to

the belief that in this department of nature at least certain "properties" had been imparted to matter and it had then been left to act largely by itself; yet, since the vital processes of living organisms are capable of exact correlation with all other forces, such as light, heat, and electricity, the direct action of this universal all-controlling Mind in all the phenomena of nature seems demonstrated beyond a doubt, leaving apparently little or no room for any action of second causes.

But this view of the matter, as is very evident, is liable to lead to a pantheistic view of the universe, than which nothing could be more horrible.

How then shall we reconcile these conflicting views?

In this case, as in so many others, the Bible comes in to show us the rational _via media_, the straight path of reason and sound philosophy which avoids the absurdities of both extremes.

The plain and unambiguous teaching of the Bible is that God, the Creator, is a being, a person, infinite in all His powers and perfections, omnipresent throughout the universe; yet that there is a place in which He is to be found, or where He abides, in a sense in which He is not to be found in any other place. This paradox is easily understood when we realize that God is present everywhere throughout His universe _by His word and by His Spirit,_ --His word being as effective throughout the remotest corners of His universe as near at hand, for the very simple reason that matter has no "properties" which He has not imparted to it, and therefore it can have no innate inertia or reluctance to act which God's word would need to overcome in order to induce it to act, even when this word operates across the boundless fields of space. He has created free personalities, and He leaves the mind of each of His creatures free to serve Him or not to serve Him, these free intelligent beings becoming thus true second causes. More than this, provision for almost innumerable second causes seems to have been made even among other departments of nature, without however interfering with the direct action of the word of the Infinite One in guiding and controlling them all.

Christ Jesus, our Lord and Savior, was associated with the Father in all the primary work of Creation; and He came to earth to show us what God the Father is like, that mortals might behold their Creator without being consumed. In Him we are to behold as much of the Deity as it is for our good to know; beyond that we must trust the hand that never wearies, the mind that never blunders, the heart that never grows cold.

In reality the seeming conflict between the doctrine of second causes and that of God's omnipresence is closely analogous to the old (imaginary) conflict between the Law and the Gospel, read from the book of nature instead of from the Bible. The reign of second causes is the reign of law; but God's immediate action brings in the supernatural, the miraculous, or the Gospel. Each has its proper place; and neither must be dwelt on to the exclusion of the other. We are all under the hard exactitude of the law, with its irrevocable condemnation, until the Gospel intervenes, and not only pardons the past, but enables us to fulfil the law's requirements for the future. The reign of second causes alone would take away man's moral responsibility, making us all mere creatures of our environment, the victims of a merciless determinism, and death would be the inevitable result of the violation of the slightest physical or physiological law. But we are all given power to live above environment, and a beneficent healing power is constantly intervening to save us from the consequences of our errors, healing our wounds and curing our diseases, in this giving us an object lesson of the forgiveness of sin and a promise of our ultimate conquest over all its power. We are all ineluctably bound about by countless chains of second causes, "awful with inevitable fates," until we see through them all the close providential working of our Creator, who is also our Saviour, and who is in no way shackled by His own laws, but conducts all things according to the counsel of His own will.

The Bible teaches us of a Creation as a definite act, completed at a definite period in the past, and it gives us the Sabbath as the divine memorial of this completed Creation. We have seen how science also points backward along the various diverging lines of the great perspective of the ages to the vanishing point whence they all begin, the birth-day of the world; and we say that thus science confirms the Bible record of Creation. But we also know that when Christ was being examined by the Sanhedrin for healing on the Sabbath, He defended Himself by saying, "My Father worketh hitherto, and I work." That is, although "the works were finished from the foundation of the world," and second causes are now largely operative in nature all around us, still there is everywhere manifest an active energy, a presence, an Intelligence, "in Whom we live, and move, and have our being."

That we cannot comprehend all this, that we cannot set definite boundaries to these seemingly conflicting views, is not at all surprising; for we are but finite.[55] Even His universe partakes so much of His prerogative of infinity that it is utterly beyond the compass of our finite minds. Indeed, if either the Bible or the book of nature contained nothing beyond what we could easily comprehend, would it not diminish our reverence and awe for the One behind them, Whom we now regard as infinite in power and in wisdom?

True, the natural human heart cannot bear this thought of the direct acting throughout nature of the infinite Creator. It brings us too close beneath His gaze in our sinful shortcoming and nakedness.

[Footnote 55: A recent clever writer likens some of these metaphysical speculations to the act of a baby sucking at a nursing bottle. So long as there is any milk in the bottle, the baby sucks with pleasure and profit. Unfortunately the little fellow does not always stop sucking when the supply of milk gives out, but still keeps on sucking empty air, with resulting discomfort and colic. We all need to recognize the limits of the intellectual milk supply, and not keep on trying to solve problems that are in their very nature beyond the limits of the human mind.]

And so men draw the veil of their pantheistic or monistic philosophy over their hearts, to hide them from His all-searching gaze. In ancient times they seem to have done the same, as the monuments of Egypt and Babylonia declare; and the intimate knowledge of Nature and its Creator which they had in the morning of our world, degenerated into the nature worship and polytheism which we find so nearly universal at the first dawn of secular history. It is only the child of God, the redeemed man, who can view without flinching the sublime fact of a direct Creation, or face the other great fact that what we call second causes are not the real causes of natural action, that the ordinary phenomena of light, heat, gravity, vital action, etc., do not occur because certain "properties" have been once imparted to matter and it then left to act of itself, any more than the child of God is left to struggle along with the supply of divine grace which was imparted to him at his conversion. The Christian feels his constant dependence upon his Creator for overcoming power day by day, and he sees the whole universe just as momently dependent upon the tireless watchcare of the great Sustainer of all. The Christian alone delights to look upon the ceaseless service of his Father's love, perpetually ministering to the needs and even to the whims of His creatures. But if this tireless ministry reminds man of his own spiritual nakedness and insular selfishness, it serves also to remind him that it is only the free gift of a righteousness not his own that can clothe the ashamed soul cowering beneath the eye of infinite Purity and unselfish Love.

In our natural state we are like the dead, inorganic matter. Only by a new life that must be imparted to us from above, a real, individual, new creation, can we become alive spiritually. And then only by constant dependence for spiritual life and growth upon the word of the One who first created us can we hope to develop into His true sons and daughters, whose continuous care is momently exercised in controlling every particle of our bodily frame, and by whose continuous guidance in the development of character we hope to become worthy of a place in His presence forevermore.

V

Our Lord Jesus once said to the leaders of the Jews, "If ye believed Moses, ye would believe me; for he wrote of me. But if ye believe not his writings, how shall ye believe my words?" (John 5: 46-47). In our days is certainly consistent and appropriate that those who have had their faith revived in the first chapters of the Bible should also have renewed confidence in the last part of the Bible. A belief in a real Creation of the world, as recorded in the book of Genesis, naturally implies a belief in the end of the world as predicted in the book of Revelation. A belief in the former destruction of the world by water is in accord with a belief in its coming destruction by fire, each of these destructions being not absolute but regenerative.

This is in fact the line of argument used in that remarkable prophecy of 2 Peter 3: 3-7:

"In the last days mockers shall come with mockery, walking after their own lusts, and saying, Where is the promise of his coming? For, from the days that the fathers fell asleep, all things continue as they were from the beginning of the creation. For this they wilfully forget, that there were heavens of old, and an earth compacted out of water and amidst water, by the word of God; by which means the world that then was, being overflowed with water, perished; but the heavens that are now, and the earth, by the same word have been stored up for fire, being reserved against the day of judgment and destruction of ungodly men."

Two points in this remarkable prophecy deserve special attention:

1. It is a description of the religio-scientific problems of the "last days"; and the class of people referred to are represented as "mocking" at the second coming of Christ, because they have grown accustomed to denying, or "wilfully forgetting," the former destruction of the world by the waters of the Flood. This prediction, as we have seen, is in complete and accurate accord with the present situation; for the doctrine of Evolution is chiefly supported by the accepted theories of geology that there never was a universal Flood. Belief in the current theories of geology and in a universal Deluge cannot be held by the same mind, for they are mutually exclusive: either one makes the other meaningless. And as the popular geology is the foundation of the Evolution theory, so does the latter render useless and incredible what the Bible calls "that blessed hope," the second coming of Christ and the purification of the

earth by fire.

2. The mockers here described certainly talk exactly like our modern _uniformitarians_; for they argue that "from the days that the fathers fell asleep, all things continue as they were from the beginning of the creation." They imply that in the days of "the fathers" some people were foolish enough to believe differently; but since they "fell asleep" we have learned better. It should also be carefully noted that their theory of uniformity stretches back, not to the close of Creation, but to "the beginning of the Creation." Plainly, then, _Creation itself is embraced in their scheme of absolute uniformity_; and according to their view all distinction is smoothed out between Creation and the present perpetuation of the world by second causes. How could we ask for a more accurate word picture of the modern popular doctrines of the evolutionists and their characteristic methods of reasoning than is here given us by an inspired prophecy nearly two thousand years ago?

VI

The call of the hour to the Church of Christ is for a renewed confidence in that Guide Book which she has brought with her down the centuries. As her Divine Lord went away, He commissioned her to carry His good tidings to all peoples; and so long as she remained true to this commission and to her instruction book, the world's cunning sophistries could not deceive her, nor could the cruel power of a world empire stifle her voice. And now when her absent Lord is about to return again, it surely behooves her to set her house in order, and to return with candor and fidelity to that written code of instruction left with her by her departing Master.

For the old-time friends of the Bible, the night of darkness and doubt is rapidly passing; the morning of a fuller knowledge and a fuller confidence is at hand. Gone are those agonies of doubt regarding the truthfulness of the Bible's history and the adequacy of its ethics for the needs of our modern world. Abandoned forever are all those futile attempts at compromise, in a vain and painful endeavor to translate the record of Creation into the language of a pseudo-science now rapidly being outgrown, and to adapt the plan of salvation to the false standards of an artificial age that seems to be rapidly disintegrating before the Church's very eyes. She now realizes that her Bible is more accurate than the world's science, her simple gospel wiser than its philosophy.

The hour has struck; a sublime opportunity is before her; for the God of nature has Himself opened up before His Church the long-sealed chapters in His larger book, and is now pointing out the marvellous agreement between His book of nature and His written record. The strongest message of the Church has often been heard amid the darkest ages of apostasy. And the prophecies of the Bible have repeatedly pointed out a special message that the Church is to bear to the world in that darkest hour just before the breaking of eternal day,--a message that we now see is wonderfully adapted to this age of evolutionism in science and pantheism in philosophy. Looking down along the darkening vistas of the coming years, the great Jehovah saw how a vastly increased knowledge of His created works would be perverted into a burlesque of Creation, and how this would result in a wide-spread apostasy in which His written Word would be derided and scorned. Thus He timed a special reform for His faithful people to give to the world just before the end, calling upon the disbelievers in Creation then living to "worship him that made heaven, and earth, and the sea, and the fountains of waters" (Rev. 14: 7). And so now, when the darkness of evolutionism and pantheism is most dense, a light from above has illuminated the record in the book of nature, the language of which is already more familiar to our modern world than the language of the book so long distrusted and almost derided. This message itself from the book of nature is full of the essential ideas of the Gospel, faith in a Creator, who by His tireless care for the particles composing our bodies keeps them in order, and by healing our injuries and curing our diseases inspires us with faith in Him as our Saviour and Redeemer. And in such an hour, in such a world crisis, He has placed within the power of His Church these modern means of travel and quick communication, in order to speed on this last work of His Church so as to complete it in "this generation."

* * * * *